TRANSNATIONAL NETWORKS AND CROSS-RELIGIOUS EXCHANGE IN THE SEVENTEENTH-CENTURY MEDITERRANEAN AND ATLANTIC WORLDS

W0018709

Universal Reform: Studies in Intellectual History, 1550–1700

Series Editors
Howard Hotson, St Anne's College, Oxford,
Vladimír Urbánek, Academy of Sciences of the Czech Republic, Prague

The fission of the western Church in the Reformation era released great quantities of energy, not all of which was contained by the confessional churches. Alongside the well-studied process of confessionalisation and the persistence of variety within popular religion, the post-Reformation period witnessed a series of poorly understood attempts by a wide variety of intellectuals to extend the reforming impulse from the spheres of church and theology to many different areas of life and thought. Within these ambitious reforming projects, impulses originating in the Reformation mixed inextricably with projects emerging from the late-Renaissance and with the ongoing transformations of communications, education, art, literature, science, medicine and philosophy. Although specialised literatures exist to study these individual developments, they do not comfortably accommodate studies of how these components were sometimes brought together in the service of wider reforms. By providing a natural home for fresh research uncomfortably accommodated within Renaissance studies, Reformation studies and the histories of science, medicine, philosophy and education, this new series will pursue a more synoptic understanding of individuals, movements and networks pursuing further and more general reform by bringing together studies rooted in all of these sub-disciplinary historiographies but constrained by none of them.

Transnational Networks and Cross-Religious Exchange in the Seventeenth-Century Mediterranean and Atlantic Worlds

Sabbatai Sevi and the Lost Tribes of Israel

BRANDON MARRIOTT

Routledge
Taylor & Francis Group

LONDON AND NEW YORK

First published in paperback 2024

First published 2015 by Ashgate Publishing

Published 2016 by Routledge
4 Park Square, Milton Park, Abingdon, Oxon OX14 4RN

and by Routledge
605 Third Avenue, New York, NY 10158

Routledge is an imprint of the Taylor & Francis Group, an informa business

British Library Cataloguing in Publication Data
A catalogue record for this book is available from the British Library.

The Library of Congress has cataloged the printed edition as follows:
Marriott, Brandon, 1982– author.
    Transnational Networks and Cross-Religious Exchange in the Seventeenth-Century Mediterranean and Atlantic Worlds : Sabbatai Sevi and the Lost Tribes of Israel / by Brandon Marriott.
        pages   cm. – (Universal Reform, Studies in Intellectual History, 1550–1700)
    Includes bibliographical references and index.
    1. Shabbethai Tzevi, 1626-1676. 2. Naylor, James, 1617?-1660. 3. Apocalyptic literature – History and criticism. 4. Transmission of texts – History – 17th century. 5. Lost tribes of Israel. I. Title.
    BM755.S45M37 2015
    201'.509032–dc23   2014048290

ISBN: 978-1-4724-3584-2 (hbk)
ISBN: 978-1-03-292871-5 (pbk)
ISBN: 978-1-315-54980-4 (ebk)

DOI: 10.4324/9781315549804

To Alexander, this book put you to sleep when you were fussy

Contents

List of Figure

Acknowledgements

The completion of this ambitious project would not have been possible without the guidance and thoughtful comments of Howard Hotson, Luke Clossey, Nicholas Guyatt, Nicholas Davidson and a host of other academics whose formal and informal advice, suggestions for readings and research, and help with languages and translations were essential in bringing it together. These scholars include Vladimir Urbanek, Leigh Penman, Joanna Weinberg, Lyndal Roper, Judith Pfeiffer, Stefano Villani, Mario Infelise, Charles Gehring, Ida Toth and Andrew Redden, as well as the anonymous reviewers of this text. Alongside the staff at the Warburg Institute where this book came to its completion, I would like to thank all of the archivists and librarians in England, Israel, Italy, the Netherlands and the United States who helped while I bumbled my way through the research that was generously funded by grants from the Gerda Henkel Stiftung, the Social Sciences and Humanities Research Council of Canada, the Rothschild Foundation of Europe and the Spalding Trust.

To my spouse, Laura Dalby, who has heard more about eschatology than I am sure she ever wanted, I am truly grateful. As important was the familial support of Ron, Gwen and Darren Marriott. Finally, I would like to thank my friends who proofread this manuscript willingly or were held captive in the car as I read it aloud to them, housed me during research trips or met me in different cities, and suggested conferences and books. Roger Doxey, Ryan Sumal, Michael Bangloy, Pheroze Unwalla, Keian Noori, Chelsea Sutcliffe, Agnieszka Nowicka, Kamila Cwiklinska, Thomas Maring, Richard Anderson, Sukhjit Chohan, Alexander Janos, Angelika Teutsch, Sebastian Reinstadler and Ruben Leavitt, I have appreciated your help and distraction.

List of Abbreviations

AAS	American Antiquarian Society, Worcester
ASV	Archivio di Stato di Venezia (Venetian State Archives), Venice
BNCF	Biblioteca Nazionale Centrale di Firenze (the Central National Library of Florence), Florence
BR	Bibliotheca Rosenthaliana in the University of Amsterdam Library, Amsterdam
BRBL	Beinecke Rare Books and Manuscripts Library at the University of Yale, New Haven
JTS	The Jewish Theological Seminary, New York
LSF	London Society of Friends' Library, London
MHS	Massachusetts Historical Society, Boston
NA	Nationaal Archief (National Archives), The Hague
NYPL	New York Public Library, New York
SA	Archivio Segreto Vaticano (Vatican Secret Archives), Vatican City
SP	State Papers in the National Archives, London
TNA	The National Archives, London
ULL	University Library of Leiden, Leiden

Introduction

In 1669 CE, the Protestant scholar Petrus Serrarius set out on a journey from the Dutch Republic to the Ottoman Empire.[1] Like many European Christians travelling similar routes, Serrarius was on a religious mission. Yet he was not on pilgrimage to Jerusalem like most. He went in another direction, heading for the home of the Ottoman *divan* (or court) in Adrianople to meet Sabbatai Sevi, the Jewish messiah who apostatised to Islam and became one of the sultan's gatekeepers. At a time when long-distance travel entailed much hardship, why was a Dutch Christian crossing an entire continent to see a Jewish convert to Islam he had neither met nor corresponded with before?

Serrarius was born into a Dutch merchant family in London. Growing up in seventeenth-century England, he studied at Christ Church in Oxford before returning to his fatherland, where he obtained a master's degree in theology at the University of Leiden. Serrarius was a 'pious man, but also a learned one' and, after completing his education, he embarked upon a career in pastoral ministry.[2]

A short time later, Serrarius moved to Amsterdam, where he befriended the Jewish rabbi Menasseh ben Israel.[3] A small group developed around these two men, and Serrarius and his Christian friends began working with Menasseh and other Dutch Jews on projects relating to Jewish–Christian reconciliation. Serrarius' close relationships with members of the *Sephardim* (the Jews of Spanish and Portuguese descent) meant that he was frequently informed of the latest news circulating among the Dutch Jewry. When the letters announcing the messiahship of Sabbatai Sevi were received by the Amsterdam Jewry, Serrarius was one of the first Christians to hear about them.

---

[1]   All dates in this text are CE unless otherwise stated.

[2]   Ernestine van der Wall, *De Mystieke Chiliast Petrus Serrarius (1600–1669) en Zijn Wereld* (Leiden: I.G.C. Printing, 1987), 624. For more on Serrarius, see the aforementioned book as well as Ernestine van der Wall, 'The Amsterdam Millenarian Petrus Serrarius (1600–1669) and the Anglo-Dutch Circle of Philo-Judaists', in Johannes van den Berg and E.G. van der Wall, eds, *Jewish–Christian Relations in the Seventeenth Century: Studies and Documents* (Dordrecht: Kluwer Academic Publishers, 1988), 73–94.

[3]   For more on Menasseh ben Israel, see Yosef Kaplan, Henry Mechoulan and Richard Popkin, eds, *Menasseh ben Israel and His World* (Leiden: E.J. Brill, 1989); David Katz, 'Menasseh ben Israel's Mission to Queen Christina of Sweden, 1651–1655', *Jewish Social Studies* Vol. 45, No. 1 (1983), 57–72; Cecil Roth, *A Life of Menasseh ben Israel: Rabbi, Printer, and Diplomat* (Philadelphia: The Jewish Publication Society of America, 1945); Benjamin Shmidt, 'The Hope of the Netherlands: Menasseh ben Israel and the Dutch Idea of America', in Paolo Bernardini and Norman Fiering, eds, *The Jews and the Expansion of Europe to the West, 1450 to 1800* (New York and Oxford: Berghahn Books, 2001), 86–106; and Ernestine van der Wall, 'Three Letters by Menasseh ben Israel to John Durie: English Philo-Judaism and the Spes Israelis', *Nederlands Archief voor Kerkgeschiedenis* Vol. 65 (1985), 46–63.

Sabbatai Sevi was born and raised a continent away from Serrarius. Sabbatai spent the years of his youth in the Ottoman port town of Smyrna, which was home to diverse populations of Jews, Christians and Muslims. While Sabbatai received a thorough religious training at a traditional Jewish educational institution known as a *yeshiva*, it was mysticism in the form of the *kabbalah* that truly fascinated him. One day, in 1648, Sabbatai claimed to hear the voice of God say to him: 'Thou art the saviour of Israel, the messiah'.[4] From that moment onwards, Sabbatai started committing strange acts, including pronouncing the ineffable name of God.

Sabbatai's actions did not go unnoticed, and the heads of the Jewish community convened and banished him from Smyrna. Forced out of his hometown, Sabbatai wandered throughout the Ottoman Empire, travelling to Istanbul, Salonika and eventually Jerusalem. Perhaps it was on the streets of the old city that a young man named Nathan Benjamin Levi first saw him.

Nathan was an exceptionally gifted student, competent in both Jewish theology and rabbinical law. When Nathan finished his schooling in Jerusalem, he married and settled in Gaza, where he began to study the kabbalah. Then, in early 1665, Nathan claimed to have a prophetic vision in which it was revealed to him that Sabbatai Sevi was the true messiah.

Nathan and Sabbatai met in Gaza shortly thereafter. A fervent discussion ensued and Sabbatai publicly proclaimed himself the messiah. Nathan was respected by many Jews in Gaza, and they accepted the authenticity of his vision and spread news of it to their friends and close acquaintances via correspondence and word of mouth. With Nathan conducting an extensive written proselytising campaign, Sabbatai went from one Jewish community to the next, gaining more and more followers. Within a year, Sabbatai was the most popular Jewish messiah since Jesus; his name was known from Yemen to the West Indies.

The growing messianic movement came to the attention of the Ottoman authorities, who responded by ordering the arrest of the Jews' new saviour. While Sabbatai's arrest and subsequent imprisonment were unexpected, they did not diminish the Jews' enthusiasm. Visitors flocked to his prison cell from across Europe and the Levant, bribing the guards simply to kneel before their messiah. Excitement was at its peak when Sabbatai was finally called before the Ottoman divan to face the consequences of his actions. His Jewish followers expected the sultan to be so awestruck by Sabbatai that he would willingly yield his crown and become the messiah's servant.

This meeting, however, was to surprise and dismay Sabbatai and those who believed in him. Standing in front of the highest members of the Ottoman government,

---

[4]   Sabbatai apparently shared this information with rabbi Solomon Laniado in Aleppo when he met him in 1665. Laniado's letter containing this story is published in Gershom Scholem, *Sabbatai Sevi: The Mystical Messiah*, trans. R.J. Zwi Werblowsky (Princeton: Princeton University Press, 1976), 136. Sabbatai's early messianic claims are discussed more thoroughly in Chapter 2, 'New Monarchs or Grand Impostors? James Nayler and Sabbatai Sevi (1656–1666)'. For more on Sabbatai Sevi, Nathan of Gaza and the history of Sabbatianism, see the sources listed in notes 13 and 14 as well as Chapter 4, 'A Jewish Messiah among Christians: The Evolution of European Perceptions of Sabbatai Sevi (1665–1666)'.

Sabbatai was given an ultimatum. The choice, he was told, was simple: convert to Islam or die. With the executioner preparing to make him a martyr, Sabbatai relented and converted to Islam. As a reward, he was given a post at the court and a pension. Thus, Sabbatai walked out of the interview not with the sultan as his servant, but as the new Muslim servant of the sultan.

The messiah's apostasy sent shock waves across the Jewish world, and the majority of Jews abandoned the Sabbatian movement as quickly as they had joined it. Although the disarray of his followers was indescribable, the messianic excitement had been so great that the apostasy could not quash it entirely. Sabbatai himself vacillated in his behaviour. Sometimes he would act as a pious Muslim and revile Judaism. At other times, he would write letters to Jewish communities in which he continued to sign his name, 'Messiah of the God of Israel and Judah, Sabbatai Sevi'.[5] Some Sabbatian believers even copied Sabbatai's example and embraced Islam too, founding the *Donmeh* – a small sect of Jewish converts to Islam who dissimulated their messianic beliefs.[6]

When the initial news of the apostasy reached Amsterdam, the Dutch scholar Petrus Serrarius chose to support the few Jews who remained faithful to the messiah. Three years later, Serrarius set out in hope of meeting Sabbatai himself. While the Dutch Protestant died en route to the Levant, never meeting the man who had captivated his attention for the last few years, Serrarius' journey points to a paradox central to this study. In an era of heightened confessionalisation,[7] why was a Christian exhibiting such loyalty to a Jewish convert to Islam?

This was a period in which the proliferation of views about the coming end often fanned the flames of inter-confessional conflict. Members of other faiths were frequently portrayed in negative apocalyptic terms. Some Protestants believed that

---

[5]   Ben Zvi Institute facsimile of MS 2262/79: Sabbatai Sevi to the Jewish community of Berat, August 1676.

[6]   For more on the Donmeh, see Marc Baer, *The Donme Jewish Converts, Muslim Revolutionaries, and Secular Turks* (Stanford: Stanford University Press, 2009); Marc Baer, 'Globalization, Cosmopolitanism, and the Donme in Ottoman Salonica and Turkish Istanbul', *Journal of World History* Vol. 18, No. 2 (2007), 141–70; and the chapter entitled 'The Crypto-Jewish Sect of the Donmeh (Sabbatians) in Turkey', in Gershom Scholem, *The Messianic Idea in Judaism: And Other Essays on Jewish Spirituality* (London: Schocken Books, 1971).

[7]   According to the confessionalisation thesis, this was a time in which confessional walls were erected, children were indoctrinated and religious identities were hardened through education, theological training and the establishment of customs. See William Monter, 'Religion and Cultural Exchange, 1400–1700: Twenty-First-Century Implications', in Heinz Schilling, Istvan Toth and Robert Muchembled, eds, *Cultural Exchange in Early Modern Europe I* (Cambridge: Cambridge University Press, 2006), 9. For more on confessionalisation in early modern Europe, see John Hedley, Hans Hillerbrand and Anthony Papalas, eds, *Confessionalization in Europe, 1555–1700: Essays in Honor and Memory of Bodo Nischan* (Aldershot: Ashgate, 2004); Wim Janse and Barbara Pitkin, *The Formation of Clerical and Confessional Identities in Early Modern Europe* (Leiden: E.J. Brill, 2006); Benjamin Kaplan, *Divided by Faith: Religious Conflict and the Practice of Toleration in Early Modern Europe* (London and Cambridge, Mass.: The Belknap Press of Harvard University Press, 2007); and Ronnie Hsia, *Social Discipline in the Reformation: Central Europe 1550–1750* (London: Routledge, 1991).

the pope was the antichrist and the Ottoman Empire in the Levant was Gog and Magog, the evil forces of the antichrist who would appear in the last days.[8]

Meanwhile, certain Jews hoped that their messiah would save them from their Christian oppressors and return them to the promised land currently under Muslim occupation.[9]

Alongside this mutual antipathy was another process – a process in which increased cross-religious interactions facilitated reconciliation by providing a framework for people of different faiths to come together, share their ideas and formulate new expectations that blurred traditional religious boundaries. For instance, the Protestant view of the Catholic Church as the ultimate apocalyptic enemy coupled with its 'prejudice and ignorance' towards Islam opened the way for the growth of Protestant philo-Semitic interpretations of the last days. In other words, the Jews became the natural allies of the Protestants against their two greatest threats: the pope in Rome and the sultan in Istanbul.[10]

It is the lesser-known process of cross-religious transmission and fusion that this study investigates.[11] This project is ultimately one that uses four case studies to demonstrate the interconnectedness of apocalyptic thinking across continents and confessional divides between 1648 and 1666.[12] The narrative begins in 1648 with a crypto-Jewish Converso named Antonio de Montezinos, who claimed to have discovered the Lost Tribes of Israel hidden in the jungles of South America. Montezinos' story, originating and travelling along Jewish networks in the 1640s, was

---

[8]   Richard Cogley, 'The Fall of the Ottoman Empire and the Restoration of Israel in the "Judeo-Centric" Strand of Puritan Millenarianism', *Church History* Vol. 72, No. 2 (2003), 318–19.

[9]   Abraham Gross, 'The Expulsion and the Search for the Ten Tribes', *Judaism* Vol. 41, No. 2 (1992), 145.

[10]   For more on Judeocentric Puritan eschatology, see Cogley, 'The Fall of the Ottoman Empire and the Restoration of Israel in the "Judeo-Centric" Strand of Puritan Millenarianism', 304–32.

[11]   Because the term 'syncretism' has problematic implications, this study follows Patrick Manning in understanding this phenomenon in terms of 'fusion' – the combining of pre-existing and new ideas into a new cultural creation. For more on this concept, see Patrick Manning, *Navigating World History: Historians Create a Global Past* (New York: Palgrave Macmillan, 2003), 281.

[12]   In regard to the terminology associated with apocalyptic thinking, terms with the roots 'millenarian' or 'millennial', 'chiliast', 'apocalyptic' and 'eschatology' pertain to religious beliefs about the end of history even though there is no academic agreement about what any of these words mean. Richard Cogley, 'Seventeenth Century English Millenarianism', *Religion* Vol. 17 (1987), 393, takes a broader definition and uses 'millenarian' to refer to any seventeenth-century Englishman who 'used the last things as historiographical categories through which he derived the full intelligibility of history' because more restrictive definitions 'fail to comprehend the diversity of opinion and the range of activity that recent scholarship has brought to light'. They also direct away from the common objective of people who believed that the last things were unfolding in time. As such, this study follows Cogley in using these terms broadly, but it focuses on the shared nature of terms such as 'eschatological' that can refer to movements of all three Abrahamic religions. More specifically, it tends to use 'messianic' to refer to Jewish movements based on the expectation of a messiah and 'millennial' to Christian movements that often expected the imminent thousand-year reign of Jesus. For more on definitions, see Cogley, 'Seventeenth Century English Millenarianism', 379–96.

shared with Protestants throughout northern Europe and in the Americas who used it to argue that the American indigenous peoples were the descendants of the Israelites.

From the Atlantic world, the focus of this text shifts to Europe and the Levant. The second chapter centres on the Quaker messiah James Nayler, who rode into Bristol in 1656 in a manner that replicated Jesus' entry into Jerusalem. Reports about the Bristol affair were dispersed throughout England, the Dutch Republic and the Italian peninsula in letters, gazettes, newsbooks and pamphlets. Although the chapter proves that Europeans knew about Nayler to a degree not previously acknowledged, it also examines the possible networks along which the story of Nayler could have been brought to the Ottoman Empire and argues against certain historiographical claims that it informed the messianic career of Sabbatai Sevi.

The final two chapters explore the westward transmission of news and rumours associated with the outbreak of the Jewish Sabbatian movement. The third chapter presents the history of the sack of Mecca in 1665. While the majority of this text discusses real people, accounts based in heavily documented testimony and intellectual theories, the tale of Mecca's destruction was completely fictitious. Even though it was only a rumour, it was in many ways similar to the other narratives: it connected Jews, Christians and Muslims from the Middle East through Europe to North America; it was written about and printed in correspondence and publications in a variety of languages; and it affected the ways in which its readers understood their own religion as well as that of others. The last chapter looks at Christian responses to the Jewish messiah in handwritten mercantile and diplomatic dispatches as well as printed gazettes and pamphlets from 1666.

The first purpose of this study is to tell this story – to reveal, explore and analyse the process in which these eschatological ideas emerged, were passed on from one religious community to another and were transformed through their transmission. Doing so would be impossible without a series of pre-existing scholarly traditions.

There is a wealth of material on Sabbatai Sevi and his followers. Gershom Scholem's *Sabbatai Sevi: The Mystical Messiah* (1976) has been supplemented by many other studies that explore different elements of the Sabbatian movement.[13] There is also a rich Hebrew historiography on Sabbatianism, headlined by a two-volume collection edited by Rachel Elior entitled *The Sabbatian Movement and its Aftermath: Messianism, Sabbatianism and Frankism* (2001).[14] Alongside the

---

[13] These include Ada Rapoport-Albert, *Women and the Messianic Heresy of Sabbatai Zevi 1666–1816* (Oxford: Littman Library of Jewish Civilization, 2011); Matt Goldish, *The Sabbatean Prophets* (Cambridge, Mass.: Harvard University Press, 2004); Elisheva Carlebach, *The Pursuit of Heresy: Rabbi Moses Hagiz and the Sabbatian Controversies* (New York: Columbia University Press, 1990); Jane Hathaway, 'The Grand Vizier and the False Messiah: The Sabbatai Sevi Controversy and Ottoman Reform in Egypt', *Journal of the American Oriental Society* Vol. 117, No. 4 (1997), 665–71; Moshe Idel, 'On Prophecy and Magic in Sabbateanism', *Kabbalah: Journal for the Study of Jewish Mystical Text* Vol. 8 (2003), 7–50; and a host of other articles, including those by Jacob Barnai, who has written extensively on Sabbatai's Jewish and Converso followers in the Ottoman Empire. For more, see the select bibliography at the end of this book.

[14] Rachel Elior, *The Sabbatian Movement and its Aftermath: Messianism, Sabbatianism and Frankism* (Jerusalem: Institute of Jewish Studies, 2001). Other Hebrew texts include Gershom Scholem, ed., *Studies*

specific academic articles and books, much research has been undertaken on Jewish messianism more generally.[15]

In reference to the study of seventeenth-century Christian millennial and messianic movements under consideration in this text, almost all of the 20 books about James Nayler limit their discussion to his English environment.[16] A broader approach has been used in the study of larger groups, such as the Fifth Monarchy Men and the Puritans, who were situated throughout the Atlantic world and are therefore often examined within this regional arena.[17]

Turning to the eschatological beliefs in the Lost Tribes, Tudor Parfitt's *The Lost Tribes of Israel: The History of a Myth* (2002) and Zvi Ben-Dor Benite's *The Ten Lost Tribes: A World History* (2009) present the entire history of the ancient Hebrews across the Judeo-Christian world from biblical times to the modern age. In regard to the specific stories about the Lost Tribes discussed herein, Montezinos' account has garnered formidable research whereas the rumour of the sack of Mecca has received scant attention.[18]

---

and Texts Concerning the History of Sabbatianism and its Metamorpheses (Jerusalem: Mossad Bialik, 1982) and Yehuda Liebes, ed., *G. Scholem, Researches in Sabbatianism* (Tel Aviv: AmOved, 1991). Hebrew articles include Isaiah Sonne, 'New Material on Sabbatai Zevi from a Notebook of R. Abraham Rovigo', *Sefunot* Vol. 3, No. 4 (1960), 39–70; Jacob Barnai, 'A Document from Smyrna Concerning the History of Sabbatianism', *Jerusalem Studies in Jewish Thought* Vol. 2 (1982), 118–31; and Yosef Kaplan, 'The Attitude of the Leadership of the Portuguese Community in Amsterdam to the Sabbatian Movement', *Zion* Vol. 39 (1974), 198–216, among others.

[15]   Such as Yehuda Liebes, *Studies in Jewish Myth and Jewish Messianism*, trans. Batya Stein (New York: State University of New York Press, 1993); Marc Saperstein, ed., *Essential Papers on Messianic Movements and Personalities in Jewish History* (New York and London: New York University Press, 1992); and Scholem, *The Messianic Idea in Judaism*.

[16]   For example, see William Bittle, *James Nayler 1618–1660: The Quaker Indicted by Parliament* (York: William Session Ltd., 1986); Leo Damrosch, *The Sorrows of the Quaker Jesus: James Nayler and the Puritan Crackdown on the Free Spirit* (Cambridge, Mass.: Harvard University Press, 1996); Emilia Fogelklou, *James Nayler: The Rebel Saint 1618–1660*, trans. Lajla Yapp (London: Ernest Benn Ltd., 1931); Vera Massey, *The Clouded Quaker Star James Nayler, 1618–1660* (York: Sessions Book Trust, 1999); and David Neelon, *James Nayler: Revolutionary to Prophet* (Becket: Leading Press, 2009).

[17]   Studies of seventeenth-century Puritan and Fifth Monarchist eschatology across the Atlantic world include James Holstun, *A Rational Millennium: Puritan Utopias of Seventeenth-Century England and America* (New York and Oxford: Oxford University Press, 1987); Avihu Zakai, *Exile and Kingdom: History and Apocalypse in the Puritan Migration to America* (Cambridge: Cambridge University Press, 1992); and James Maclear, 'New England and the Fifth Monarchy: The Quest for the Millennium in Early American Puritanism', *The William and Mary Quarterly* Vol. 32, No. 2 (1975), 223–60.

[18]   For more on Montezinos' story, see Ronnie Perelis, '"These Indians Are Jews!": Lost Tribes, Crypto-Jews, and Jewish Self-Fashioning in Antonio de Montezinos's Relacion of 1644', in Richard Kagan and Philip Morgan, eds, *Atlantic Diasporas: Jews, Conversos, and Crypto-Jews in the Age of Mercantilism, 1500–1800* (Baltimore: The Johns Hopkins University Press, 2009), 195–212 and the introduction to Henry Mechoulan and Gerard Nahon, *Menasseh ben Israel, The Hope of Israel: The English Translation by Moses Wall, 1652*, trans. Richenda George (Oxford: Oxford University Press, 1987).

These historiographies have been bridged by studies of cross-religious interactions that explore small groups of people in specific cities or countries.[19] But some of the statements about cross-religious influences in these works need to be reconsidered.[20]

Engaging all of these historiographies, this text tracks multiple, overlapping cultures of elite transnational apocalypticism. Be they Jewish, Christian or Muslim, the main protagonists in this study were mostly scholars and theologians or professional merchants and diplomats who not only believed that this news was important, but were also able to afford the expense of sending correspondence and publishing pamphlets about them.

While these were the men and women who wrote much of the evidence that remains, they were not indicative of the broader societies of which they were part, and this study does not seek to generalise from these small groups of people. That is not to say that only European theologians and intellectuals were interested in Sabbatai Sevi. After all, many works about the Jewish messiah were written in order to inform larger populations; however, the responses of such people are harder to gauge because they have rarely survived.

The local cultures of apocalyptic consumption based around the sermon and mostly rooted in national histories and politics intersected in multiple ways with the transnational networks of apocalyptic intelligence maintained by intellectuals and professionals. On one hand, local sermons provided the necessary background knowledge to understand the stories from the Levant that appeared in European newspapers and pamphlets. On the other hand, this foreign news sometimes prompted pastors to preach sermons that discussed the prophecies associated with the Jews and ancient Hebrews. Although much of the material spread in the elite circles dovetailed with the messages that millenarians preached to their congregation from their pulpits,

---

[19]   For English interest in Sabbatianism, see Michael McKeon, 'Sabbatai Sevi in England', *Journal of the Association for Jewish Studies Review* Vol. 2 (1977), 131–69; Michael Heyd, 'The "Jewish Quaker": Christian Perceptions of Sabbatai Zevi as an Enthusiast', in Allison Coudert and Jeffrey Shoulson, eds, *Hebraica Veritas?: Christian Hebraists and the Study of Judaism in Early Modern Europe* (Philadelphia: University of Pennsylvania Press, 2004), 234–65; and Richard Popkin, 'Three English Tellings of the Sabbatai Zevi Story', *Jewish History* Vol. 8, Nos. 1–2 (1994), 43–54. For Dutch interest, see Jetteke van Wijk, 'The Rise and Fall of Shabbatai Zevi as Reflected in Contemporary Press Reports', *Studia Rosenthaliana* Vol. 33 (1999), 7–27; N.H. van Wijk, 'Wachtend op. de Wolk naar Jeruzalem: De Verslaglegging Rond Shabbatai Tsvi in Nederlandse Pamfletten en Couranten' (doctoral thesis, University of Amsterdam, 1996); and Yosef Kaplan's introduction to the Hebrew edition of Thomas Coenen's *Vain Hopes of the Jews as Revealed in the Figure of Sabbetai Zevi* (Jerusalem: The Ben-Zion Dinur Institute for Research in Jewish History, 1998). For more on the connections between Conversos and the Sabbatian movement, see Jacob Barnai, 'Christian Messianism and the Portuguese Marranos: The Emergence of Sabbateanism in Smyrna', *Jewish History* Vol. 7, No. 2 (1993), 119–26.

[20]   This is especially true in regard to certain statements made by Richard Popkin. While Popkin was a formidable scholar who did groundbreaking work, some of his claims need to be revisited, including those in his 'Jewish–Christian Relations in the Sixteenth and Seventeenth Centuries: The Conception of the Messiah', *Jewish History* Vol. 6, Nos. 1–2 (1992), 163–77 and his 'Christian Interest and Concerns about Sabbatai Zevi', in Matt Goldish and Richard Popkin, eds, *Millenarianism and Messianism in Early Modern European Culture I* (Dordrecht: Kluwer Academic Publishers, 2001), 91–101.

the average believers may have been more likely swayed by local sermons that spoke of the political upheavals in their immediate surroundings than by stories of a Jewish messiah a continent away.

This text does not take on the whole seventeenth-century apocalyptic tradition. It neither seeks to cast aside nor re-tell the history of the well-known messianic and millenarian movements, such as those associated with the English civil wars, that are usually located within a national framework. It also does not seek to challenge the well-established Hebrew historiography on Sabbatianism from which it draws too little. Instead, it hopes to add to the historiography by showing the interconnectedness of such thinking based on evidence drawn from across continents and confessional boundaries.

This project builds on the growing body of literature in the global history of ideas by highlighting the ways in which apocalyptic ideas circulated far and wide, connecting seventeenth-century Jews, Christians and Muslims from the Levant across Europe to the Americas. In particular, it responds to calls by historians for the study of history from a larger perspective.

In terms of the Mediterranean world, it follows the scholars who have moved away from a perspective that separates it into 'East/West, Muslim/Christian, Venetian/Turk, Europe/Other' and have sought to develop a more sophisticated model that stresses the porous frontiers of the Mediterranean.[21] These historians have emphasised the permeable and fluid nature of religious landscapes where religious identities could and did change.[22] One forward-thinking historian, in particular, has called for the Sabbatian movement to be understood against a background in which Europe and the Ottoman Empire are seen as part of a shared world.[23]

At the same time, it follows the scholars of Atlantic history who have claimed that 'the Atlantic is no longer enough' because such a regional viewpoint separates northern Europe and its Atlantic colonies from the rest of the world whereas a focus on networks across a broader geographical area shows that the rest of the world was not only connected to Europe, but played an important role in shaping it as well.[24]

---

[21]   Eric Dursteler, *Venetians in Constantinople: Nation, Identity, and Coexistence in the Early Modern Mediterranean* (Baltimore: The Johns Hopkins University Press, 2006), 6, 152, 185: see Suraiya Faroqhi, *The Ottoman Empire and the World Around It* (London: I.B. Tauris, 2004); Molly Greene, *A Shared World: Christians and Muslims in the Early Modern Mediterranean* (Princeton: Princeton University Press, 2000); and David Abulafia, *The Great Sea: A Human History of the Mediterranean* (London: Penguin Books, 2011).

[22]   Zur Shalev, 'Islam, Eastern Christianity, and Superstition according to Some Early Modern English Observers', in Asaph Ben-Tov, Yaacov Deutsch and Tamar Herzig, eds, *Knowledge and Religion in Early Modern Europe* (Leiden: Brill, 2013), 139.

[23]   Jacob Barnai, 'Some Social Aspects of the Polemics between Sabbatians and their Opponents', in Matt Goldish and Richard Popkin, eds, *Millenarianism and Messianism in Early Modern European Culture I* (Dordrecht: Kluwer Academic Publishers, 2001), 82.

[24]   David Armitage and Michael Braddick, eds, 'Introduction', *The British Atlantic World, 1500–1800* (New York: Palgrave Macmillan, 2009), 9. As Adam Sutcliffe has warned, 'Atlantic Jewish history, if not placed within a global context, risks further reinforcing this misleading and unhelpful division' between the westward-expanding European states and the Muslim lands. For more, see Adam Sutcliffe, 'Jewish History

It is from this vantage point that this project tracks the thread of eschatological transmission that began in the Americas in the 1640s, spread across Europe to the Levant in the 1650s and returned to the Americas via Europe in the 1660s.

Because of its cross-religious approach, this study uses a fresh body of sources. Alongside well-documented letters, pamphlets and gazettes discussed by numerous historians, it presents printed Italian *avvisi* (or gazettes) from Genoa and Venice, a Dutch newsbook published in Haarlem, and English mercantile and diplomatic correspondence written across Europe that all refer to the Jewish messiah but have not appeared in the historiography. Similarly, the story of the Quaker messiah James Nayler was found in locations that historians have not expected, and this study has unearthed printed Italian avvisi from Milan and Genoa as well as Dutch pamphlets and newsbooks from Haarlem and Amsterdam that spread the account of Nayler's messianic entrance into Bristol to a wider European audience. Piecing these fresh sources together with well-known material provides greater clarity into the conceptualisation of cross-religious and transnational transmission of eschatological constructs as well as the resulting beliefs that blurred traditional religious boundaries across the seventeenth-century Abrahamic world.

In order to explore the movement of information across both the Atlantic and Mediterranean worlds, this study employs the concept of networks. Networks appear to be a useful tool in the research of intellectual communication because they highlight aspects of human society that are otherwise neglected.[25] Indeed, networks are a 'useful heuristic device' because they serve to identify particular channels through which individuals, goods and knowledge moved.[26]

The networks vital for the circulation of narratives about the messiah and the Lost Tribes of Israel between 1648 and 1666 were fluid, multifaceted and often overlapped.

---

in an Age of Atlanticism', in Richard Kagan and Philip Morgan, eds, *Atlantic Diasporas: Jews, Conversos, and Crypto-Jews in the Age of Mercantilism, 1500–1800* (Baltimore: The Johns Hopkins University Press, 2009), 18–30, 25.

[25] There are patterns of relations crucial in the flows of information, influence, goods and contagious diseases that hardly anyone could imagine before they were discovered and mapped with networks. For more, see Jeroen Bruggeman, *Social Networks: An Introduction* (London and New York: Routledge, 2008), 2.

[26] Francesca Trivellato, 'Sephardic Merchants in the Early Modern Atlantic and Beyond: Towards a Comparative Historical Approach to Business Cooperation', in Richard L. Kagan and Philip D. Morgan, eds, *Atlantic Diasporas: Jews, Conversos, and Crypto-Jews in the Age of Mercantilism, 1500–1800* (Baltimore: The Johns Hopkins University Press, 2009), 119. This project follows historians who utilise networks 'in a loose sense' to highlight 'forms of affiliation and association that are less defined than a "structure" but more than a collection of individuals engaging in transactions'. These networks are 'organisations with voluntary and reciprocal patterns of exchange' that bound people together across national boundaries, nurturing and reflecting a sense of shared commitment and purpose. This definition was first used in Frederick Cooper, 'Networks, Moral Discourse and History', in Thomas Callaghy, Ronald Kassimir and Robert Latham, eds, *Intervention and Transnationalism in Africa* (Cambridge: Cambridge University Press, 2001), 24 and has subsequently been employed in Gary Magee and Andrew Thompson, eds, *Empires and Globalisation: Networks of People, Goods and Capital in the British World, c. 1850–1914* (Cambridge: Cambridge University Press, 2010), 27.

Because presenting them in a simplified, formulaic manner would misrepresent them, it might be best to limit an introductory discussion of the networks to their origins.

Many of the networks grew out of, or at least expanded greatly because of, coinciding waves of migration in the early modern period. These population displacements spread people to new locations where they retained ties to their dispersed friends and family, creating transnational networks that were then used to move more people, goods and information – all of which reinforced the connections.[27] At the same time, some of the events that led to the creation of the networks were interpreted in a manner that heightened eschatological expectations.

To begin, the Jewish expulsion from Spain in 1492 spread the 200,000 Sephardic Jews throughout the Mediterranean world. Many of them travelled the long distance to the multi-ethnic and multi-religious Ottoman Empire where they were welcomed for their business skills and knowledge of the Iberian Peninsula.[28] While large populations settled in the political and economic capital of Istanbul (formerly Constantinople) as well as the emerging port city of Smyrna,[29] other exiles stayed in the Italian peninsula. In Venice, the Jews could only live in the walled-in ghetto where they paid a third higher rent,[30] whereas the authorities in Tuscany actually solicited the immigration of Jews to Livorno to encourage the growth of an international port that would rival those in Venice and Genoa.[31]

---

[27]   Transnational ethnic and religious connections often offered a more secure way of building or expanding economic networks because compatriots were viewed as more trustworthy; members of family and ethnic groups could co-operate to overcome logistical challenges, to obtain finances, to co-ordinate transportation and to regulate transactions. For more, see Ghislaine Lydon, *On Trans-Saharan Trails: Islamic Law, Trade Networks, and Cross-Cultural Exchange in Nineteenth-Century Western Africa* (Cambridge: Cambridge University Press, 2009), 343.

[28]   Minna Rozen, 'Strangers in a Strange Land: The Extraterritorial Status of Jews in Italy and the Ottoman Empire in the Sixteenth to the Eighteenth Centuries', in Aron Rodrigue, ed., *Ottoman and Turkish Jewry: Community and Leadership* (Bloomington: Indiana University Turkish Studies, 1992), 125.

[29]   In Istanbul, the Jewry of 3000 households before 1492 had grown almost threefold by 1535. Meanwhile, Smyrna was a town of little consequence at the time of the expulsion, and the Sephardim originally avoided it until a sixteenth-century economic crisis in Salonika's textile trade led many Jews to resettle in Smyrna, which was growing in economic importance due to the Ottoman–Venetian war that had obstructed the sea route to Istanbul. Third and fourth generation descendants of the exiles, therefore, decided to go to the Anatolian port city, where they formed wealthy, trade-oriented communities. See Jonathan Israel, *European Jewry in the Age of Mercantilism 1550–1750* (Oxford: The Littman Library of Jewish Civilization, 1998), 23; Daniel Goffman, *Izmir and the Levantine World, 1550–1650* (Seattle and London: University of Washington Press, 1990), 79, 82, 90; Jacob Barnai, 'The Sabbatean Movement in Smyrna: The Social Background', in Menachem Mor, ed. *Jewish Sects, Religious Movements, and Political Parties: Proceedings of the Third Annual Symposium of the Philip M. and Ethel Klutznick Chair in Jewish Civilization held on Sunday–Monday, October 14–15, 1990* (Omaha: Creighton University Press, 1992), 119.

[30]   Brian Pullan and David Chambers, eds, *Venice: A Documentary History, 1450–1630* (Oxford: Blackwell, 1993), 338–9, 344.

[31]   Francesca Trivellato, *The Familiarity of Strangers: The Sephardic Diaspora, Livorno, and Cross-Cultural Trade in the Early Modern Period* (New Haven and London: Yale University Press, 2009), 77.

The Spanish expulsion was so traumatic that certain Jews interpreted it as the beginning of the 'birth pangs of the messiah', the disturbances which would occur immediately prior to the coming of the messiah and their ultimate redemption.[32] For 40 years after the expulsion, there was a deep messianic yearning among some Jewish scholars, including Abraham ben Eliezer ha-Levi, Isaac Abravanel and Samuel Usque. These heightened messianic expectations manifested themselves in numerous ways: contemporary events were given greater eschatological significance, people began prophesying and calling for prayer vigils to hasten the coming of the messiah, and mystical texts, such as the *Zohar*, were disseminated more broadly.[33]

As important, the expulsion inaugurated a shift in the central location of the kabbalah from Iberia to the Levant. According to at least one scholar, there 'can be no doubt that the expulsion of the Jews from Spain and Portugal at the end of the 15th century was *the* decisive event that changed the course of the history of Kabbalah in North Africa, as it did the development of Kabbalah in general'.[34] The sixteenth-century kabbalist Isaac Luria, in particular, provided a solution to the problem of exile by tying the cosmic ideas of exile and redemption to the historic struggle of the Jews. Luria, who moved to Safed in 1569, was aware of the trauma brought about by the expulsion and argued that exile was punishment for sin, a test of faith and part of the mystical mission in which the Jews worked towards their redemption. Luria's 'great myth of exile' gave meaning for the suffering of the Sephardim by connecting their physical experience to the supernatural exile of man from God, which apparently created a messianic tension not previously found in the kabbalah.[35]

Whether or not one believes that it was the possibility of messianism perpetuated through Lurianic Kabbalism that became explicit a century later in the Sabbatian movement,[36] or that Jewish messianic activity from the fifteenth to the eighteenth century, culminating in the Sabbatian movement, was 'a direct response to the expulsion of the Jews from Spain in 1492',[37] it seems certain that the Spanish

---

[32] The biblical passage of Isaiah 66:7–9 is probably the source of this expression. Liebes, *Studies in Jewish Myth and Jewish Messianism*, 40.

[33] Scholem, *The Messianic Idea in Judaism*, 37, 41–2.

[34] Moshe Idel, 'Jewish Mysticism Among the Jews of Arab/Moslem Lands', *The Journal for the Study of Sephardic and Mizrahi Jewry* Vol. 1, No. 1 (2007), 22.

[35] W.D. Davies, 'From Schweitzer to Scholem: Reflections on Sabbatai Svi', *Essential Papers on Messianic Movements and Personalities in Jewish History* (New York and London: New York University Press, 1992), 345; Gershom Scholem, 'Isaac Luria: A Central Figure in Jewish Mysticism', *Bulletin of the American Academy of Arts and Sciences* Vol. 29, No. 8 (1976), 8–13; Gershom Scholem, *Major Trends in Jewish Mysticism* (London: Thames and Hudson, 1955), 286.

[36] David Biale, 'Gershom Scholem on Jewish Messianism', *Essential Papers on Messianic Movements and Personalities in Jewish History* (New York and London: New York University Press, 1992), 531; Scholem, *Major Trends in Jewish Mysticism*, 246, 288.

[37] David Ruderman, 'Hope Against Hope: Jewish and Christian Messianic Expectations in the Late Middle Ages', *Exile and Diaspora: Studies in the History of the Jewish People* (Jerusalem: Ben-Zvi Institute, 1991), 187. While revisionists have argued that the effects of the expulsion on Jewish mysticism as well as the influence of Lurianic Kabbalism on Sabbatianism have been over-emphasised, there are still important connections that should not be completely discounted. For the revisionist argument, see the work of

expulsion laid the foundation of the transnational Sephardic networks and promoted messianism among some Jews.

Returning to the Iberian Peninsula, 120,000 Spanish exiles went to Portugal, where they were forcefully converted to Christianity five years later when Portugal sought to create a uniform religious identity. With this one political manoeuvre, the entire Jewish population became *Conversos* (Christian converts of Jewish origin).[38] Many of the Conversos fled Portugal, especially following the increased attention of the Portuguese Inquisition after 1547. Approximately 50,000 Conversos travelled to the Ottoman Empire, where they too were welcomed for their linguistic skills, familial networks and knowledge of the Iberian Peninsula. A particularly large Converso population moved to Smyrna, where they returned to Judaism and established two of the most prominent and richest congregations.[39] In their new home, they retained connections to their brethren throughout Europe and were known to frequent Istanbul, Venice and Amsterdam.[40]

While the Conversos who settled in Venice were expelled in 1550 because they were believed to be malevolent, faithless people who infected Christians with 'a wicked and evil doctrine',[41] the Conversos flocked to Livorno after the Tuscan state issued a charter in 1593 that made the Jews of Livorno the first Jewish community in the Christian world without religious restrictions. Returning to Judaism, the Conversos helped to turn the port city into the main Mediterranean trade hub between the Levant and northern Europe.[42]

Moshe Idel, especially his '"One from a Town, Two from a Clan": The Diffusion of Lurianic Kabbalah and Sabbateanism – a Re-examination', *Jewish History* Vol. 7, No. 2 (1993), 79–104 and the subsection entitled 'Messianism and Eschatology' in his 'Religion, Thought, and Attitudes: The Impact of the Expulsion on the Jews', in Elie Kedourie, ed., *Spain and the Jews: The Sephardi Experience, 1492 and After* (London: Thames and Hudson, 1992), 123–39.

[38]   Renata Segre, 'Sephardic Refugees in Ferrara: Two Notable Families', in Benjamin Gampel, ed., *Crisis and Creativity in the Sephardic World 1391–1648* (New York: Columbia University Press, 1997), 165: Conversos have been called *Marranos*, New Christians and crypto-Jews. While many were forcefully converted, some did so voluntarily. The religious beliefs of Conversos were very complex: some were practising Christians, others were only Christians externally and practised Judaism in secret, and many more mixed Christianity and Judaism in unique ways. For more on the Conversos, see David Graizbord, *Souls in Dispute: Converso Identities in Iberia and the Jewish Diaspora, 1580–1700* (Philadelphia: University of Pennsylvania Press, 2004) and Yirmiyahu Yovel, *The Other Within, The Marranos: Split Identity and Emerging Modernity* (Princeton and Oxford: Princeton University Press, 2009).

[39]   Barnai, 'The Sabbatean Movement in Smyrna', 114–15.

[40]   Barnai, 'Christian Messianism and the Portuguese Marranos', 121, 122.

[41]   Pullan and Chambers, eds, *Venice: A Documentary History, 1450–1630*, 345. Conversos who wanted to stay in Venice often left for a short period of time to the Ottoman Empire and then returned as a *mercante levantino*, a Levantine Jewish merchant. A brief sojourn in the Empire was all that was needed to launder 'past sins' because Ottoman Jewish subjects residing in Venice were entitled to freedom of trade, freedom of movement and tax exemptions that their co-religionists in the Republic were not. See Rozen, 'Strangers in a Strange Land', 126, 129.

[42]   Lionel Kochan, *The Making of Western Jewry, 1600–1819* (New York: Palgrave Macmillan, 2004), 28–9. Livorno's growth was noticed in Venice, a state currently in economic decline due to the Ottoman–Venetian war, and the Venetian council met and reconsidered its policy: it now sought to

Because of their conversions to Christianity, the Conversos were accepted in countries where the Jews were not. The regents of Amsterdam permitted the settlement of the Conversos for their economic abilities, 'trusting that they were Christians' even though the leaders of the Reformed Church opposed their admission.[43] Seeing the immense freedom of many of the Conversos who returned to Judaism in Amsterdam, many Sephardic Jews moved to the Dutch Republic, where they too engaged in trade.[44]

A final group of Conversos fled westward across the Atlantic and joined the Jews from the Dutch Republic and England who founded the first overt Jewish settlements in the western hemisphere.[45] The American communities were one end of the complex social, familial, ethno-religious and mercantile networks of Converso and Sephardic Jews that stretched from Brazil through England, the Dutch Republic and Italy to the Ottoman Empire. No other diaspora ranged as widely, linked as many empires or cut across as many confessional divides as that of the Sephardim.[46]

Like the Spanish expulsion, the forced conversions in Portugal in 1497 were also interpreted as part of the messianic birth pangs. Some thought that the event in Portugal was more tragic and thus a more prominent sign of the imminent end.[47]

The importance of the Conversos would come to the fore in both pre- and post-apostasy Sabbatianism. In Sabbatai's youth, several of his close friends and fellow students were from Converso families, and they had a significant impact on Sabbatai's messianic interests.[48] Later, with the widespread outbreak of Sabbatianism, the centres of messianic enthusiasm were linked to the geographic distribution of the Iberian

---

also attract Jewish and Converso merchants to compete with Tuscany. For more on the Sephardim in Livorno, see Trivellato, *The Familiarity of Strangers*; Olivia Remie Constable, *Housing the Stranger in the Mediterranean World: Lodging, Trade, and Travel in Late Antiquity and the Middle Ages* (Cambridge: Cambridge University Press, 2003); and Benjamin Arbel, *Trading Nations: Jews and Venetians in the Early Modern Eastern Mediterranean* (Leiden: E.J. Brill, 1995).

[43]   Maarten Prak, *The Dutch Republic in the Seventeenth Century: The Golden Age* (Cambridge: Cambridge University Press, 2005), 216; Thomas Glick, 'On Converso and Marrano Ethnicity', *Crisis and Creativity in the Sephardic World 1391–1648* (New York: Columbia University Press, 1997), 70.

[44]   The Jewish traders in Amsterdam not only owned a quarter of the shares of the Dutch East India Company, but they also utilised their connections in the American colonies to expand the economic influence of the Dutch West India Company. See Jan de Vries and Ad van der Woude, *The First Modern Economy: Success, Failure, and Perseverance of the Dutch Economy, 1500–1815* (Cambridge: Cambridge University Press, 1997), 152.

[45]   Jacob Marcus, *The Colonial American Jew 1492–1776* (Detroit: Wayne State University Press, 1970), 1328.

[46]   In some ways, the Sephardim and Iberian crypto-Jews actually constituted two distinct networks. Yet there were enduring links between them, consisting of family ties, religious sympathies and active trading collaboration. For more on this, see Jonathan Israel, *Diasporas within a Diaspora: Jews, Crypto-Jews and the World Maritime Empires* (Leiden: E.J. Brill, 2002).

[47]   Isaiah Tishby, 'Acute Apocalyptic Messianism', in Robert Seltzer, Frank Fitzgerald and Marc Saperstein, eds, *Essential Papers on Messianic Movements and Personalities in Jewish History* (New York and London: New York University Press, 1992), 263; Yosef Hayim Yerushalmi, *From Spanish Court to Italian Ghetto: Isaac Cardoso: A Study in Seventeenth-Century Marranism and Jewish Apologetics* (Seattle and London: University of Washington Press, 1981), 304–5.

[48]   Barnai, 'The Sabbatean Movement in Smyrna', 116.

exiles and, in particular, of former Conversos.[49] These communities received the Sabbatian gospel eagerly because it struck a chord with them: they, their parents or their grandparents had been forced to live lives of dissimulation in Iberia.[50] After the apostasy, when the majority of Sabbatai's Jewish followers abandoned their faith in him, Sabbatianism still found a particularly strong response among former Conversos because their own experience prepared them to accept the rationalisation of the messiah's conversion: it was just an outward mask necessary to cover a different inner experience. Sabbatai was living a life similar to that of a crypto-Jewish Converso.[51] The forced conversions in Portugal, therefore, also helped to set the geographical and intellectual stage for the Sabbatian movement.

In Christendom, the reforming aspirations created antagonisms that gave way to confessionalisation and violence that forced communities to displace themselves.[52] Petrus Serrarius' family, for example, left the Low Countries after the Spanish invasion and settled in London, where they used their contacts to build a mercantile network across the channel.

Although the English authorities accepted the Dutch refugees, they persecuted their own Puritan population who dispersed throughout the Atlantic world. The scattering of Puritans from New England to the West Indies enabled them to form wide-ranging mercantile networks that were reinforced by blood relationships and family credit arrangements.[53] Puritans often sent letters along these networks that began with a religious exhortation followed by business updates, cash reckonings, news, opinions and even rumours.[54]

The Puritans too understood their experience in an eschatological manner: some saw New England as a refuge designed by God for his chosen people to escape the imminent judgement that would soon fall on the rest of the world. Such 'eschatological expectations and apocalyptic visions constituted the very theme of the Pilgrims' migration to Plymouth, and later they constituted the very motive of the great Puritan migration to New England'.[55]

---

[49]   Stephen Sharot, *Messianism, Mysticism, and Magic: A Sociological Analysis of Jewish Religious Movements* (Chapel Hill: The University of North Carolina Press, 1982), 101.

[50]   Scholem, *Sabbatai Sevi*, 486.

[51]   Goldish, *The Sabbatean Prophets*, 49, 46; Yerushalmi, *From Spanish Court to Italian Ghetto: Isaac Cardoso*, 304.

[52]   For more on early modern European refugees, see Heiko Oberman, '"Europa Afflicta:" The Reformation of the Refugees', *Archiv für Reformationsgeschichte* Vol. 83 (1992), 91–111.

[53]   Bernard Bailyn, *New England Merchants in the Seventeenth Century* (Cambridge, Mass.: Harvard University Press, 1955), 35, 87, 88, 144.

[54]   David Cressy, *Coming Over: Migration and Communication Between England and New England in the Seventeenth Century* (Cambridge: Cambridge University Press, 1987), 191, 178–80, 222.

[55]   In particular, the Puritans' reasoning was drawn from the prophetic vision in Revelation of the woman's flight into the wilderness in the face of the dragon's rage. This story, coupled with Thomas Brightman's correlation of England with the biblical Laodicea, encouraged thousands of Puritans to flee the doomed 'English Laodicea' to New England in the 1630s. See Zakai, *Exile and Kingdom*, 130; and Maclear, 'New England and the Fifth Monarchy', 227.

While migrants escaping religious intolerance often went to their new homes for economic incentives, these centuries also witnessed the expansion of foreign trade with the integration of commerce across Europe, the Mediterranean and the Atlantic. In the sixteenth century, English and Dutch merchants began bypassing their traditional Italian intermediaries and established trade networks that linked the Ottoman Empire to northern Europe via the Italian peninsula.[56] Officially formed in 1581, the English Levant Company grew rapidly and employed 40,000 people by 1605,[57] including 25 factors in Istanbul. By 1649, Smyrna had overtaken Istanbul as the largest English colony, with over 50 merchants. Dutch shipping soon matched pace and, between 1645 and 1648, there were approximately 97 Dutch ships a year sailing between the Dutch Republic and the Ottoman Empire.[58]

While these merchants developed their commercial operations in the Mediterranean, their counterparts in the Atlantic broke the Iberian monopoly on transatlantic trade and set up colonies in the Americas.[59] Overlapping participation in both regions was common, from employees and investors who worked for and backed joint stock companies in both the Mediterranean and the Atlantic to adventurers and sailors who brought maritime practices across this artificial divide.[60]

These merchants often installed representatives, sometimes family, abroad whose correspondence provided information that gave them an edge over their competitors.[61] Merchant letters, which 'served as sinews holding together European commerce',[62] primarily contained discussions of economic news, markets, exchange rates and descriptions of products, as well as comments on political, religious and

---

[56] Italian dominance was finally undermined in the seventeenth century due to the discovery of new sea routes to Asia, the rise of Ottoman and Hapsburg powers, the Thirty Years' War, the bankruptcy of Spain and the restrictions that the sixteenth-century sultans placed on the Italians to squeeze them out of trade in Ottoman lands. It is difficult to identify anything as clear-cut as 'Muslim', 'French' or 'Christian' trade. While the literature refers to the English and French as coherent communities, it was more like states struggling to impose a national trade policy on a disparate collection of individuals. For more, see Molly Greene, 'Beyond the Northern Invasion: The Mediterranean in the Seventeenth Century', *Past and Present* Vol. 174, No. 1 (2002), 42–71.

[57] Nabil Matar, *Islam in Britain, 1558–1685* (Cambridge: Cambridge University Press, 1998), 10.

[58] De Vries and van der Woude, *The First Modern Economy*, 381.

[59] Nuala Zahedieh, 'Making Mercantilism Work: London Merchants and Atlantic Trade in the Seventeenth Century', *Transactions of the Royal Historical Society* Vol. 9 (1999), 144. For more on various merchant communities in the Atlantic, see Peter Coclanis, ed., *The Atlantic Economy during the Seventeenth and Eighteenth Centuries: Organization, Operation, Practice, and Personnel* (Columbia: University of South Carolina Press, 2005).

[60] Lauren Benton, 'The British Atlantic in Global Context', in David Armitage and Michael Braddick, eds, *The British Atlantic World, 1500–1800* (New York: Palgrave Macmillan, 2009), 277–8, 288.

[61] Jan Willem Veluwekamp, 'International Business Communication Patterns in the Dutch Commercial System, 1500–1800', in Hans Cools, Marika Keblusek and Badelock Noldus, eds, *Your Humble Servant: Agents in Early Modern Europe* (Hilversum: Uitgeverij Verloren, 2006), 122, 127–8.

[62] Francesca Trivellato, 'Merchants' Letters across Geographical and Social Boundaries', in Francisco Bethencourt and Florike Egmond, eds, *Cultural Exchange in Early Modern Europe III* (Cambridge: Cambridge University Press, 2007), 10–23.

military affairs that were believed to have an economic impact.[63] These were usually less than four pages, sent weekly or at least regularly, and presented news that was generally considered reliable. Increased trade, therefore, gave rise to the widespread movement of mercantile news between the Levant, the Americas and Europe.

Professional diplomatic networks were also established in this period, with the installation of permanent ambassadors at foreign capitals.[64] While many European states, including England, the Dutch Republic and Venice, had extensive diplomatic networks by 1650, the Ottoman Empire did not. The sultans refused to treat the European powers on 'a basis of equality and reciprocity',[65] choosing to rely instead on their own extraordinary ambassadors as well as European diplomats stationed in Istanbul for their foreign relations.

Wherever they were located, diplomats wrote regular reports about their host state based on their own observations, news acquired by trading items of political importance and statements from informants who were cultivated through special favours and bribery. These dispatches, which were sent to their home governments, were similar to merchant letters in many regards. They were handwritten, usually only a few pages, sent regularly and contained news that was considered reliable.[66]

Scientific institutes, such as the Royal Society in England, similarly established vast intellectual networks.[67] Henry Oldenburg, the society's secretary, employed the mercantile and diplomatic networks as well as connections that he built himself to collect information from distant lands for dissemination among the academic members of the society. Oldenburg, for instance, sent his letter to Persia through Sir Andrew Riccard, the chairman of the East India and Levant Companies, and he asked John Finch, the English consul in Florence, to find out about the latest scientific ventures of the Italian intellectuals.[68]

At the same time, the advent of news industries brought about more networks that facilitated the widespread dissemination of information. Starting in Italy, but

---

[63]   Trivellato, 'Merchants' Letters across Geographical and Social Boundaries', 88.

[64]   For more on diplomacy, see Donald Queller, *The Office of Ambassador in the Middle Ages* (Princeton: Princeton University Press, 1967); Keith Hamilton and Richard Langhorne, *The Practice of Diplomacy: Its Evolution, Theory and Administration* (London and New York: Routledge, 1995); and Garrett Mattingly, *Renaissance Diplomacy* (Baltimore: Penguin Books, 1964).

[65]   Edhem Eldem, 'Foreigners at the Threshold of Felicity: The Reception of Foreigners in Ottoman Istanbul', in Donatella Calabi and Stephen Christensen, eds, *Cultural Exchange in Early Modern Europe II* (Cambridge: Cambridge University Press, 2007), 120.

[66]   Mario Infelise, 'News Networks between Italy and Europe', *The Dissemination of News and the Emergence of Contemporaneity in Early Modern Europe* (Farnham: Ashgate, 2010), 52.

[67]   For more on the Royal Society, see Michael Hunter, *Establishing the New Science: The Experience of the Early Royal Society* (Woodbridge: The Boydell Press, 1989) and Michael Hunter, *The Royal Society and its Fellows 1660–1700: The Morphology of an Early Scientific Institution* (Preston: Alphaprint, 1982).

[68]   Henry Oldenburg to John Finch, 7 December 1665 and 10 April 1666, as quoted in Rupert Hall and Marie Boas Hall, eds, *The Correspondence of Henry Oldenburg III* (Madison, Milwaukee and London: The University of Wisconsin Press, 1965–73), 86, 632–4. For an overview of the life and work of Henry Oldenburg, see Marie Boas Hall, *Henry Oldenburg: Shaping the Royal Society* (Oxford: Oxford University Press, 2002).

quickly spreading across Europe to England, handwritten and then printed news sources appeared with regularity in Venice, Amsterdam and London.[69]

Written in the vernacular to reach the broadest possible audience, Italian *avvisi* were often a page or two in length, with each paragraph presenting a different news item from another place.[70] The handwritten and printed *avvisi* were circulated throughout the city, the state and beyond.[71] The news re-exported from Venice was generally considered reliable abroad and, because of its geographical position and longstanding ties to the Levant, reports from the Ottoman Empire often first appeared in the Venetian *avvisi*.[72]

Following the Venetians, the Dutch developed their own news sources. In fact, the first Dutch *coranto* began with a piece of news from Venice, and each subsequent issue resembled the avvisi in content and format.[73] By the middle of the seventeenth century, corantos had been replaced by pamphlets and gazettes that supplied the Dutch population with the latest news. While pamphlets were filled with military, political and religious reports as well as sensational stories, gazettes were seen to carry more sober information supplied by respectable compatriots on the scene, such as ambassadors, clergymen, soldiers and merchants.[74] Two prominent Dutch gazettes that were known for their accurate and objective reporting were the *Oprechte Haerlemse Saterdaegse Courant* and the *Oprechte Haerlemse Dingsdaegse Courant*. A Dutch newsbook entitled the *Hollandtze Mercurius* was also published weekly, but with consecutive page numbers so at the end of the year it could be put together to form a book.[75]

---

[69] For more on the development of the news industry, see Filippo de Vivo, *Information and Communication in Venice: Rethinking Early Modern Politics* (Oxford: Oxford University Press, 2007); Craig Harline, *Pamphlets, Printing, and Political Culture in the Early Dutch Republic* (Dordrecht: Martinus Nijhoff Publishers, 1987); Otto Lankhorst, 'Newspapers in the Netherlands in the Seventeenth Century', in Sabrina Baron and Brendan Dooley, eds, *The Politics of Information in Early Modern Europe* (London: Routledge, 2001), 151–9; and Joad Raymond, *The Invention of Newspaper: English Newsbooks 1641–1649* (Oxford: Clarendon Press, 2005).

[70] Zsuzsa Barbarics and Renate Pieper, 'Handwritten Newsletters as a Means of Communication in Early Modern Europe', in Francisco Bethencourt and Florike Egmond, eds, *Cultural Exchange in Early Modern Europe III* (Cambridge: Cambridge University Press, 2007), 60, 55.

[71] Infelise, 'News Networks between Italy and Europe', 66, 55.

[72] Mario Infelise, 'From Merchants' Letters to Handwritten Political Avvisi: Notes on the Origins of Public Information', in Livio Antonielli, Carlo Capra and Mario Infelise, eds, *Cultural Exchange in Early Modern Europe III* (Cambridge: Cambridge University Press, 2007), 36–7. For more on the development of the news industry in Venice, see de Vivo, *Information and Communication in Venice*; Mario Infelise, 'Copisti e Gazzettieri nella Venezia del Seicento', in S. Gasparri, G. Levi and P. Moro, eds, *Venezia: Itinerari per la Storia della Citta* (Bologna: Societa Editrice il Mulino, 1997), 193–219; Mario Infelise, *Prima dei Giornali: Alle Origini della Pubblica Informazione* (Bari: Laterza, 2002); and Mario Infelise, 'Sulle Prime Gazette a Stampa Veneziane', in Livio Antonielli, Carlo Capra and Mario Infelise, eds, *Per Marino Berengo: Studi degli Allievi* (Milano: FrancoAngeli, 2000), 469–89.

[73] De Vivo, *Information and Communication in Venice*, 81.

[74] Van Wijk, 'The Rise and Fall of Shabbatai Zevi as Reflected in Contemporary Press Reports', 15, 17.

[75] For more on the Dutch news industry, see Craig Harline, *Pamphlets, Printing, and Political Culture in the Early Dutch Republic* (Dordrecht: Martinus Nijhoff Publishers, 1987) and Otto Lankhorst,

Originally, England had lagged behind the rest of Europe in developing periodical news. By the 1660s, however, English gazettes, newsbooks, newsletters and pamphlets were constantly being printed and circulated.[76] The official *Gazette*, first released in 1665, was published twice a week and sold up to 15,000 copies per issue. It had the potential to reach an audience of two or three times more because it was recopied, passed among groups of friends and read on market days in cities and towns around England.[77] Sometimes these publications were even sent across the Atlantic. By 1648, John Winthrop had received 13 newsbooks from England,[78] and he thanked John Davenport 'for the 2 weekely Intelligences' in 1666.[79]

None of these networks existed in isolation: information often flowed among and across them. While such transmission constitutes the heart of this study, these networks serve as essential background because they allowed for the vast movement of eschatological ideas between 1648 and 1666. In sum, these networks were formed through population displacement beginning as early as 1492 that spread people throughout the Abrahamic world and contributed to the expansion of apocalyptic thought, forming both the geographical and intellectual context for the four interconnected case studies discussed forthwith.

'Newspapers in the Netherlands in the Seventeenth Century', in Sabrina Baron and Brendan Dooley, eds, *The Politics of Information in Early Modern Europe* (London: Routledge, 2001), 151–9.

[76] Nicholas Brownlees, 'Narrating Contemporaneity: Text and Structure in English News', in Brendan Dooley, ed., *The Dissemination of News and the Emergence of Contemporaneity in Early Modern Europe* (Farnham: Ashgate, 2010), 229, 245.

[77] Sonja Schultheiss-Heinz, 'Contemporaneity in 1672–1679: The Paris *Gazette*, the London *Gazette*, and the *Teutsche Kriegs-Kurier* (1672–1679)', in Brendan Dooley, ed., *The Dissemination of News and the Emergence of Contemporaneity in Early Modern Europe* (Farnham: Ashgate, 2010), 120. For more on the *London Gazette*, see P.M. Handover, *A History of the London Gazette 1665–1965* (London: Her Majesty's Stationery Office, 1965).

[78] Sabrina Baron, 'The Guises of Dissemination in Early Seventeenth-Century England', in Brendan Dooley and Sabrina Baron, eds, *The Politics of Information in Early Modern Europe* (London: Routledge, 2001), 50–51; Raymond, *The Invention of Newspaper*, 237, 252.

[79] John Davenport to John Winthrop, 20 July 1666, as quoted in Isabel MacBeath Calder, ed., *Letters of John Davenport: Puritan Divine* (New Haven: Yale University Press, 1937), 168.

Chapter 1

The Lost Tribes in the Americas: Judeo-Christian Reciprocity across the Atlantic World (1648–1666)

From the Jewes our faith began,
To the Gentiles then it ran,
To the Jewes returne it shall,
Before the dreadfull end of all.[1]

In 1492, Christopher Columbus landed in the Bahamas archipelago, and his well-known voyage marks the beginning of transatlantic travel. But Columbus also started a lesser-known intellectual trend upon meeting the American aboriginals. Thinking he had arrived on the outskirts of Asia, he misidentified them. He called them *indios*.

The Iberians who followed Columbus cared little about the aboriginals' origins. It was not until later, when Amerigo Vespucci first identified the Americas as a *Mundus Novus* that questions about the history of their inhabitants came to the fore.[2] Saint Augustine had stated that even the most monstrous races came from Adam and Eve, and the pope had concurred. The indigenous peoples were fully human and therefore had to be connected to the biblical world. To deny them such an origin was to place them outside of scripture, which would make the holy book incomplete and inadequate.[3]

The lone survivors of the great flood were Noah and his family. If the indigenous peoples were descendants of Noah, how did they travel to the Americas in a manner that fit into the accepted chronology in the book of Genesis? Numerous theories emerged that linked the aboriginals to the Phoenicians, to the Arabians or to the people of the biblical Ophir. One of the more creative answers was offered by the French millenarian Isaac La Peyrere, who postulated that they came from men who

[1]   Thomas Thorowgood, *Iewes in America, or, Probabilities that the Americans are of that Race* (London, 1650), 23–4.

[2]   For more on Vespucci, see the chapter entitled 'Vespucci's Tabloid Journalism, 1497–1504' in David Abulafia, *The Discovery of Mankind: Atlantic Encounters in the Age of Columbus* (New Haven and London: Yale University Press, 2008).

[3]   For more on the importance of this discovery and its impact on European intellectuals, see Abulafia, *The Discovery of Mankind* and Anthony Grafton, *New World, Ancient Texts: The Power of Tradition and the Shock of Discovery* (Cambridge, Mass.: The Belknap Press of Harvard University Press, 1992).

existed before Adam.[4] Yet even in his aptly titled *Prae-Adamite* (1655), La Peyrere tied his argument into the Bible. Indeed, no scholar before the seventeenth century put forth a hypothesis without biblical foundations.[5]

Eventually the question of the origins of the American indigenous peoples intersected with a separate and much older problem. Early modern scholars had never fully determined what had become of the Lost Tribes of Israel. The story of the Lost Tribes begins in the Bible with the ancient Hebrews, who had originally consisted of 12 tribes, descended from the 12 sons of Jacob. The 12 tribes were then divided into two camps. The two tribes of the southern kingdom of Judah and their descendants became the Jews who spread throughout the world, whereas the other 10 tribes were led into captivity by the Assyrian King Salmanassar and disappeared entirely from the biblical record. It was this group, the northern tribes or the 10 tribes, which became known as the Lost Tribes.

With the incorporation of the Hebrew scriptures into the Christian Bible, the Lost Tribes became enshrined in the Christian canon, and Christians too started to wonder about the whereabouts of the ancient Hebrews. Because many of the messianic promises of the Old Testament were addressed to the children of Abraham, Isaac and Jacob, it was believed that God had preserved the Lost Tribes in some distant corner of the globe from which they would emerge before the end of the world.[6]

The question of the Lost Tribes' location therefore remained in the background of theological and historical debate until the discovery of the inhabitants of the Americas provided a new answer to this old problem. Scholars began to search for evidence that proved the aboriginals were descendants of the Israelites, which became known as the Lost Tribes theory. Because most European scholars would never visit the Americas themselves, they simply compared descriptions of languages, habits and buildings in order to extrapolate upon the probability.

New England Puritans, such as the minister John Cotton, were among those who speculated on this hypothesis. The Cambridge-educated Cotton had preached millenarian doctrines to his congregation in England until he received word in 1632 that he was going to receive a summons to the Court of High Commission. Cotton reacted quickly. He packed up his belongings and left for New England before the summons arrived. Like the other 21,000 Puritans who fled England in the 1620s

---

[4]   For more on La Peyrere, see Richard Popkin, 'Millenarianism and Nationalism – A Case Study: Isaac La Peyrere', in Matt Goldish and Richard Popkin, eds, *Millenarianism and Messianism in Early Modern European Culture IV* (Dordrecht: Kluwer Academic Publishers, 2001), 77–84 and Richard Popkin, ed., *Isaac La Peyrere (1596–1676): His Life, Work and Influence* (Leiden: Brill Academic Publishers, 1987).

[5]   Lee Eldridge Huddleston, *Origins of the American Indians: European Concepts, 1492–1729* (Austin and London: University of Texas Press, 1967), 11.

[6]   For a history of the Lost Tribes, see Zvi Ben-Dor Benite, *The Ten Lost Tribes: A World History* (Oxford: Oxford University Press, 2009).

and 1630s,[7] Cotton preferred to start a new life a continent away than face possible excommunication and punishment.[8]

While Cotton delivered sermons in England and New England in which he claimed that the Puritans were living in the last days,[9] he did not think that the aboriginals were related to the Lost Tribes. Like Increase Mather, a Puritan minister involved in the governing of the Massachusetts Bay colony, Cotton thought that the ancient Hebrews were somewhere in Asia. Roger Williams, the founder of Rhode Island colony, concurred. He even labelled the indigenous population: 'the Gentiles of America'.[10]

Although believed to dwell much farther away, the Lost Tribes were still important for American Christians who expected the world to end shortly. As Cotton explained in one of his sermons on the book of Revelation, the prophecies in the Bible pointed to three great events that had to occur before Jesus' return: a mass conversion of gentiles, a mass conversion of Jews and the re-emergence of the Lost Tribes. Based on the biblical calculations of English theologians such as Thomas Brightman and John Archer, Cotton anticipated that the conversion of the Jews would happen first – sometime in the middle of the 1650s.[11]

John Eliot was a member of Cotton's congregation in Boston and he had listened to many of these sermons. Eliot, who was born in England around 1604 and came to New England in 1631, first worked at the church in Boston before taking a job as the pastor of the new church in Roxbury.[12] Eliot agreed with Cotton that the Lost Tribes were in Asia, which had important implications because Eliot became a missionary to the aboriginals in 1646. Since the indigenous peoples were believed to be gentiles and not Jews, any major missionary activity was useless because they would not convert en masse until after the Jews did.

While Eliot was originally satisfied with bringing the gospel to the 'indigenous gentiles' on a small scale, his views would evolve significantly over the next 20 years. By 1650, Eliot had changed his mind. He became an outspoken proponent of the

---

[7] David Hall, *The Faithful Shepherd: A History of the New England Ministry in the Seventeenth Century* (Williamson: University of North Carolina Press, 1972), 47, 72.

[8] Michael Winship, *Making Heretics: Militant Protestantism and Free Grace in Massachusetts, 1636–1641* (Princeton and Oxford: Princeton University Press, 2002), 29, 61, 36–7.

[9] Maclear, 'New England and the Fifth Monarchy', 227.

[10] Roger Williams, *A Key to Language of America* (London, 1643), To the Reader.

[11] John Archer, *The Personal Reign of Christ upon Earth* (London, 1642), 49, calculated that the pope's 'Roman monarchy' had begun in 360 or 366, and he believed that they should 'add the 1290 dayes, which is the time how long from this it should be before the Jews should be delivered, and it makes 1650, or 1656 years of the Lord, about which time, as some have supposed, the Israelites may be delivered, by being called to Christianity'. For more, see Archer, *The Personal Reign of Christ upon Earth*.

[12] Cotton Mather, *The Life and Death of the Reverend Mr. John Eliot, who was the First Preacher of the Gospel to the Indians in America* (London, 1694), 14. For more on John Eliot, see Richard Cogley, 'John Eliot and the Millennium', *Religion and American Culture* Vol. 1, No. 2 (1991), 227–50; Richard Cogley, 'John Eliot and the Origins of the American Indians', *Early American Literature* Vol. 21, No. 3 (1986/1987), 210–25; and Richard Cogley, *John Eliot's Mission to the Indians before King Philip's War* (Cambridge, Mass.: Harvard University Press, 1999).

Lost Tribes theory, believing that his own missionary work was helping to bring about Jesus' millennial kingdom. Within a decade, however, Eliot had returned to his original position. Then, in 1666, his interest in the Lost Tribes re-emerged and would remain with him for the rest of his life. In sum, Eliot flip-flopped on this issue multiple times between 1648 and 1666.

During this period, the Puritan missionary lived in a developing town in North America that was separated from mainstream European intellectual life by an ocean. So, in such circumstances, who or what was responsible for repeatedly influencing Eliot's beliefs?

Antonio de Montezinos and the Conversos

To understand Eliot's changing perspective, one must turn to a man with an incredible story whom he would never meet: Antonio de Montezinos, or Aaron Levi. Like Eliot, Montezinos was born around 1604. Unlike Eliot, Montezinos was an Iberian Converso.

Like other young Conversos, Montezinos fled Portugal for the West Indies, where he secretly returned to Judaism. According to Montezinos' account, he came across an aboriginal named Franciscus somewhere in the Andes Mountains in 1639. Not concerned with Franciscus at first, Montezinos continued on his journey until he was arrested in Cartagena under suspicion of Judaising and locked away by the Inquisition for the next 18 months.[13]

Montezinos stated that, even though he was imprisoned, he held firm to his secret faith and at night quietly thanked God for not making him 'a Barbarian, a Black-a-Moore, or an Indian'. But one night when he said, 'Indian', he became angry with himself and thought, 'The Hebrews are Indians!' Not sure what had come over him, Montezinos remembered Franciscus and the aboriginals in the Andes whom he had witnessed praying on a Friday evening. 'Could they have been performing a Jewish service?' he wondered.[14]

When Montezinos was released from prison in 1641, he claimed to track down Franciscus, identify himself as a Jew and ask Franciscus for help. Franciscus told Montezinos that he could only do so if the Converso promised to follow him wherever he may go. Montezinos agreed. The two men set out, hiking through the Peruvian jungles for a week, resting only on the Sabbath and eating only the maize that Franciscus carried on his back. Early one morning, they came to a river and, standing on the shore, Montezinos watched as a boat full of people with skin scorched by the sun, ornaments on their feet and linen cloths tied around their heads paddled across to see them. Upon reaching Montezinos, these mysterious people turned to him and,

---

[13]   Mechoulan and Nahon, *Menasseh ben Israel*, 75. For more on accusations against Judaisers in South America, see Irene Silverblatt, 'New Christians and New World Fears in Seventeenth-Century Peru', *Comparative Studies in Society and History* Vol. 42, No. 3 (2000), 524–46.

[14]   Menasseh ben Israel, *The Hope of Israel* (London: Livewell Chapman, 1652), 3.

apparently speaking in Hebrew, they proclaimed, 'Hear O Israel, the Lord our God is one.'[15]

Montezinos said that he and Franciscus arranged their camp and waited patiently as more of these people came across the river to see them. Every boatload, however, simply repeated the same things, including that their fathers were Abraham, Isaac, Jacob and Israel. On the third day, Montezinos grew frustrated. He wanted to know more and tried to jump into their boat as it was leaving, but he almost drowned and the occupants warned him to stay away. Franciscus would inform him further, they said. When Montezinos asked Franciscus what they meant, Franciscus told him that these were the sons of Israel who were protected by God. Every time his people had tried to attack them, no one from their war party ever returned. Concluding in his own apocalyptic manner, Franciscus continued:

> the God of those Children of Israel is the true God, that all that which is engraven upon their stones is true; that about the end of the World they shall be Lords of the world ... and those Children of Israel going forth out of their Country, shall subdue the whole World to them.[16]

Whatever actually happened in South America, Montezinos headed to Amsterdam where, according to his sworn oath, he told the heads of the Sephardic community in 1644 that he had found their lost brethren. One of the men who listened to Montezinos was the rabbi Menasseh ben Israel.

Menasseh was a Sephardic Jew whose ancestors too had lived in Portugal as professing Christians for over a century. Menasseh's own father was a Converso who had been imprisoned by the Inquisition for Judaising and, upon his release, fled to Amsterdam via La Rochelle with his family. Like many Conversos, they returned to Judaism in the Dutch Republic, and Menasseh grew up to become a rabbi and a bookseller with a printing press.[17]

Because Menasseh was a scholar, he was most likely aware of the growing support for the theory that the descendants of the ancient Hebrews were in the Americas. But he had shown no interest in it until Montezinos arrived. Now the academic discourse, which was based upon a deductive line of reasoning that utilised a comparative approach, was bolstered by an eyewitness account.

Although Montezinos had just come from South America with a tale that could hardly be believed, Menasseh endorsed his story because he saw Montezinos as part of his community. According to the rabbi, Montezinos came from a respectable family, was of 'honest and known parents, 40 years old, honest, and not ambitious'.[18] Menasseh referred to him as 'our Montezinos, being a Portingal, and a Jew of our Order', and described how he spent six months with Montezinos, how Montezinos

[15]   Ben Israel, *The Hope of Israel*, 3–4: this is Deuteronomy 6:4.
[16]   Ben Israel, *The Hope of Israel*, 4–5.
[17]   Mechoulan and Nahon, *Menasseh ben Israel*, 23, 28.
[18]   Ben Israel, *The Hope of Israel*, 18.

swore an oath testifying to the truthfulness of the account in his presence and how Montezinos repeated the same oath on his deathbed two years later.[19] Menasseh believed Montezinos' narrative to such an extent that, after reading the works of numerous scholars, he stated, 'I Returne to the relation of our Montezinos, which I prefer before the opinion of all others, as most true'.[20]

Why did Montezinos tell this story, and why did people like Menasseh accept it? Not only would such claims have caused him further problems with the Inquisition, but Montezinos also made no attempt to profit from it, accepted no alms from the Jews, and apparently even turned down an offer to go to England to share his experience with interested Protestants because he wanted to keep the location of the tribe secret.[21]

While the specific source of his narrative remains unclear, Montezinos appeared to have been affected by the continuing legacy of Converso messianism. The expectation of the messiah's coming was part of the faith of some crypto-Jewish Conversos like Montezinos because it highlighted their rejection of Jesus as the messiah, which was sometimes the only Jewish principle that they held onto while dissimulating their religious beliefs.[22]

Because the return of the Lost Tribes was expected to occur alongside the advent of the messiah, Montezinos' claims were both framed and understood in an eschatological manner. For Montezinos and Conversos like him, the Lost Tribes were expected to be a powerful military force waiting in a remote location to save them and punish the Iberians.[23]

From the belief that the Lost Tribes were beyond the mythological Sambatyon river that they could not cross until the last days to legends of Prester John and his Christian kingdom at the edge of the known world, such myths informed Montezinos' account and shaped the way that it was received and recorded by his audience.[24] It is no coincidence that Montezinos' story incorporated a large river that he could not cross (like the Sambatyon) near the edge of the known world as well as a hidden population that had never been bested in warfare. While Montezinos' tale may seem completely fantastical, it ignited a spark among both Jews and Christians.[25]

---

[19]   Ben Israel, *The Hope of Israel*, 17–18.

[20]   Ben Israel, *The Hope of Israel*, 43.

[21]   Mechoulan and Nahon, *Menasseh ben Israel*, 69.

[22]   The desire to retain their Jewishness led some Conversos to believe that the appearance of the messiah was their only hope. If the Christians were right and Jesus was the messiah then their suffering made no sense. For more, see Sharot, *Messianism, Mysticism, and Magic*, 84.

[23]   Benite, *The Ten Lost Tribes*, 161.

[24]   Perelis, "'These Indians Are Jews!'", 203.

[25]   Tudor Parfitt, *The Lost Tribes of Israel: The History of a Myth* (London: Weidenfeld and Nicolson, 2002), 79–80.

Cross-religious Fertilisation and Publication

Montezinos' testimony was first spread among the Sephardim largely by word of mouth.[26] It might have stayed solely within the Jewish world were it not for the actions of an itinerant Scotsman named John Dury. An exiled Presbyterian minister repeatedly displaced by religious and military struggles, Dury's continuous migrations made him a crucial agent in stitching the pieces of this story together.[27]

In the 1640s, Dury was in the Dutch Republic on a journey throughout Europe on behalf of the unification of the Protestant world. In Amsterdam, Dury met Menasseh, who told him about Montezinos' adventure. Dury, however, was not too concerned; he did not try to get a copy of Montezinos' account. After all, reports like that were not unfamiliar to him. Six months earlier in The Hague, a 'godly man' had told Dury about the arrival of a messenger from the Lost Tribes in Istanbul.[28]

From the Low Countries, Dury continued on his way and, in 1648, he happened to come across a Presbyterian minister in Norfolk named Thomas Thorowgood. Thorowgood had been interested in the Lost Tribes theory years earlier, but it did not preoccupy his thoughts either until he came across a pamphlet about John Eliot's recent work in New England.

In order to fund and garner support for his missionary endeavour, Eliot regularly wrote about his experiences with the aboriginals in dispatches that he sent to Edward Winslow, the agent for New England in London, who lobbied the parliament and solicited funds from the churches to 'arouse enough religious enthusiasm to overcome political and financial skepticism'.[29] Eliot's letters were a useful tool in encouraging such support, so Winslow collected, compiled and passed them along for publication in pamphlets that have become known as the Eliot tracts.[30]

One of these pamphlets, which had nothing to do with the Lost Tribes, had a profound and unintended effect on Thorowgood. It convinced the Presbyterian minister that the American aboriginals could be of Jewish origin because Eliot had quickly restored them to Christianity from centuries of 'accumulated barbarianism'.[31] Thorowgood was so moved by Eliot's correspondence that he wrote,

---

[26]   Menasseh noted when he first published it that 'the famous narrative of Aaron Levi, also known by the name of Montezinos, has been widely circulated in the last few years'. See Mechoulan and Nahon, *Menasseh ben Israel*, 61–2.

[27]   For more on John Dury, see J. Minton Batten, *John Dury: Advocate of Christian Reunion* (Chicago: The University of Chicago Press, 1944); Kenneth Gibson, 'John Dury's Apocalyptic Thought: A Reassessment', *Journal of Ecclesiastical History* Vol. 61, No. 2 (2010), 299–313; and Pierre-Olivier Léchot, *Un Christianisme 'Sans Partialité': Irénisme et Méthode Chez John Dury (v. 1600–1680)* (Paris: Honoré Champion, 2011).

[28]   Thorowgood, *Iewes in America*, An Epistolicall Discourse of Mr. Iohn Dury.

[29]   Francis Jennings, 'Goals and Functions of Puritan Missions to the Indians', *Ethnohistory* Vol. 18, No. 3 (1971), 206–7.

[30]   Cogley, *John Eliot's Mission to the Indians before King Philip's War*, 66.

[31]   Richard Cogley, '"Some Other Kinde of Being and Condition": The Controversy in Mid-Seventeenth-Century England over the Peopling of Ancient America', *Journal of the History of Ideas* Vol.

> When the glad tidings of the Gospels sounding in America by the preaching of the English arrived hither, my soule also rejoyced within me, and I remembred certaine papers that had been laid aside a long time, upon review of them, and some additions to them, they were privately communicated unto such as perswaded earnestly they might behold further light.[32]

Inspired by Eliot's mission, Thorowgood became engrossed in his own investigation of the indigenous peoples and, by 1648, he had reached the same conclusion as other scholars before him: the American aboriginals descended from the Lost Tribes.

When Thorowgood met Dury in 1648, he gave the latest version of his manuscript entitled *Iewes in America* to the Scotsman. Reading Thorowgood's work reminded Dury of the two stories about the Lost Tribes that he had heard in the Dutch Republic and, after telling Thorowgood about them, he requested a copy of Montezinos' testimony from Menasseh.[33] Menasseh complied, and Dury passed it onwards to Thorowgood. The Presbyterian minister was so taken with Montezinos' narrative that he added it to his text alongside a letter from Dury that introduced it.

Unbeknownst to Menasseh, Thorowgood used the material from the rabbi to argue in favour of the Lost Tribes theory at the same time that he provided Eliot's reports as a framework to Christianise the Americas.[34] For Thorowgood, it neither mattered that Montezinos was a Converso and Eliot was a Puritan nor that their experiences occurred in different continents among different indigenous populations. Thorowgood still combined their statements to create a coherent account with eschatological implications. If the aboriginals were of Jewish origin then their conversion, which had already begun due to Eliot's efforts, would be the expected conversion of the Jews – the penultimate event before the return of Jesus.

When Dury wrote to Menasseh for a copy of Montezinos' testimony, he asked the rabbi about the Jewish perspective on the Lost Tribes. Menasseh replied

> in two Letters, telling me [Dury] that by the occasion of the Questions which I proposed unto him concerning this adjoyned Narrative of Mr. *Antonie Monterinos*, hee to give me satisfaction, had written instead of a Letter, a Treatise, which hee shortly would publish, and whereof I should receive so many Copies as I should desire.[35]

Menasseh penned this treatise, entitled the *Hope of Israel* (1650), to put a stop to the Lost Tribes theory. Unlike most of his Protestant correspondents, Menasseh believed that God had kept the Israelites as a unique people who still followed biblical ways in the Americas, Persia, Tartary and other places. The American aboriginals had only replicated some of the Jewish culture that they had witnessed among the tribe that Montezinos had discovered; they were not the descendants of the ancient Hebrews.

68, No. 1 (2007), 53.

[32] Thorowgood, *Iewes in America*, Epistle Dedicatory.
[33] Thorowgood, *Iewes in America*, An Epistolicall Discourse of Mr. Iohn Dury.
[34] Thorowgood, *Iewes in America*, 94.
[35] Thorowgood, *Iewes in America*, An Epistolicall Discourse of Mr. Iohn Dury.

Before Menasseh published his Latin edition, he printed it in Spanish for the Sephardim under the title *Esperanca de Israel* after a verse in the book of Jeremiah. He chose '*Esperanca*' for 'Hope', knowing full well the 'explosive charge' that this word had for Converso messianism.[36]

Thus, inspired by an itinerant Scotsman, the Dutch rabbi composed a treatise that harnessed 'Christian chiliastic energies to Jewish messianic ends',[37] promoting messianism among the Sephardim and introducing the Jewish messianic doctrine into Christian circles and the republic of letters.[38] Indeed, Menasseh's writings made him more famous in the Christian world as an expert on Jewish matters than he ever was in the Jewish world itself. The *Hope of Israel*, which by the early eighteenth century had been reprinted a dozen times in over half a dozen languages, not only ranks among one of the 'most influential documents of seventeenth-century Jewish history', but was certainly one of the most 'widely read pieces of early Jewish Americana'.[39]

When Dury received the account of Montezinos' testimony from Menasseh, he shared it with Edward Winslow, who was editing a selection of John Eliot's missionary letters to publish under the title *The Glorious Progress of the Gospel* (1649). Like Thorowgood, Winslow was so excited by Montezinos' story that he added it to his text even though it had nothing to do with Eliot's work because he saw the Lost Tribes theory, bolstered by Montezinos' eyewitness testimony, as a tool to encourage excitement among the English for Eliot's missionary endeavour.

For Winslow, Menasseh's treatise proved the connection between the Lost Tribes and the American aboriginals. Before he could relate the latest missionary activities of John Eliot, he stated that two questions needed to be addressed: 'What became of the ten tribes of Israel?' and 'Where are the aboriginals in the Americas from?'[40] According to Winslow, the 'Rabbi-ben-Israel, a great Dr. of the Jewes, now living at Amsterdam' claimed that the Lost Tribes were 'certainly transported into America', where he and many other New Englanders observed that the practices of the aboriginals were very similar to those of the Jews.[41] This had Christian eschatological implications for Winslow:

> It is not less probable that these Indians should come from the Stock of Abraham, then any other nation this day known in the world: Especially considering the juncture of time wherein God hath opened their hearts to entertain the Gospel, being so nigh the very years, in which many eminent and learned Divines, have from Scripture grounds, accounting to their apprehensions foretold the conversion of the Jews.[42]

---

[36] Mechoulan and Nahon, *Menasseh ben Israel*, 67. For more on the importance of '*Esperanca*' in the messianic beliefs of the Judaising Conversos, see Yovel, *The Other Within, The Marranos*, 84.

[37] Shmidt, 'The Hope of the Netherlands', 99.

[38] Mechoulan and Nahon, *Menasseh ben Israel*, 68.

[39] Shmidt, 'The Hope of the Netherlands', 92.

[40] Edward Winslow, *The Glorious Progress of the Gospel amongst the Indians in New England* (London, 1649), The Epistle Dedicatory.

[41] Winslow, *The Glorious Progress of the Gospel*, The Epistle Dedicatory.

[42] Winslow, *The Glorious Progress of the Gospel*, The Epistle Dedicatory.

Winslow's eschatologically charged introduction was followed by the missionary letters from New England that made no mention of the Lost Tribes. They simply recounted Eliot's successes, failures and needs in his work among the aboriginals. The imminent apocalypse returned to the fore in the conclusion, where Dury provided his own conjecture as to why the indigenous peoples were descendants of the ancient Israelites: 'the Jews of the Netherlands (being intreated thereunto) informe that after much inquiry they found some of the ten Tribes to be in America'.[43] This was important because:

> The palpable and present acts of providence, doe more then hint the approach of Jesus Christ: And the Generall consent of many judicious, and godly Divines, doth induce considering minds to beleeve, that the conversion of the Jewes is at hand. Its the expectation of some of the wisest Jewes now living, that about the year 1650. *Either we Christians shall be Mosaick, or else that themselves Jewes shall be Christians.* The serious consideration of the preceding Letters, induceth me to think, that there may be at least a remnant of the Generation of Jacob in America.[44]

In this manner, the accounts of a Puritan missionary in North America and a Converso in South America were brought together in northern Europe through the interaction of a Jewish rabbi and his philo-Semitic Protestant friends in England. While the former used Montezinos' narrative to propound a Jewish messianic ideology that argued the Lost Tribes were a distinct people in the Americas, the latter combined Eliot's and Montezinos' reports to advance the Lost Tribes theory, which they used as proof of the imminent Christian millennium.

The Rise of Millenarianism on Both Sides of the Atlantic

In New England, John Eliot learnt about the upsurge of interest in the Lost Tribes theory due to his continued correspondence with Edward Winslow about his missionary work. In late 1648 or early 1649, Winslow wrote to Eliot that he had embraced the belief that the aboriginals were the descendants of the ancient Hebrews. Eliot replied in 1649 that he had already started to examine the origins of the indigenous people himself; however, he could not accept the proposition that the Israelites came to the Americas. Like John Cotton, he still thought that the Lost Tribes were in Asia.[45]

Regardless, Eliot was curious to learn more and asked Winslow 'for that opinion of *Rabbi-ben-Israel* which you mention ... to know his grounds, and how he came to

---

43   Winslow, *The Glorious Progress of the Gospel*, 23.
44   Winslow, *The Glorious Progress of the Gospel*, 22.
45   Cogley, 'John Eliot and the Origins of the American Indians', 216.

that Intelligence'.[46] His friend graciously agreed and provided Eliot with the treatises of Menasseh and Thorowgood before they were even published. Reading these texts changed Eliot's mind. He himself wrote to Thorowgood:

> By reading your book, intituled, *Jews in America, or Probabilities that the Americans are of that Race*, the Lord did put it into my heart to search into some Scriptures about that subject, and by comparing one thing with another, I thought, I saw some ground to conceive, that some of the Ten Tribes might be scattered even thus far, into these parts of America.[47]

Eliot confessed to Winslow that the aboriginals' Hebrew origin placed his own work in a new perspective and confirmed his faith in the imminent millennium. In other words, the Puritan missionary accepted the apocalyptic reinterpretation of his own activities by people who would never travel to the Americas themselves. Dwelling on the 'Valley of Dry Bones' in the book of Ezekiel, Eliot believed that the dead bones that were miraculously restored to life by the preaching of God's word referred to the 'New World Jews' who would return to their covenant with God by converting to Christianity.[48] Cotton had claimed that the conversion of the Jews would begin in the mid-1650s and, if the aboriginals were Jews and not gentiles, their time of mass conversion was closer than expected.

Eliot sought out corroborating evidence for his new belief and found it in a secondhand report from a Mr. Dudley, who said that Captain Cromwell, a man recently deceased in Boston, had spoken of aboriginals in the south who were circumcised. According to Eliot, Cromwell's report was one of the most probable arguments that he had heard for the Lost Tribes theory. Eliot was so excited that he initiated a correspondence with Thorowgood who, fascinated with Eliot's direct knowledge of the indigenous peoples, requested a statement of the Puritan missionary's views. Eliot happily provided one, and Thorowgood published it under the title of 'The Learned Conjectures of the Reverend John Eliot Touching the Americans' in the revised edition of *Jews in America* (1660).[49]

Eliot's millenarian beliefs, grounded in the Lost Tribes theory, were bolstered by the ongoing military and political events in England. In particular, Eliot's understanding of the execution of Charles I in 1649 transformed him into a 'politically radical

---

[46] See the letter from John Eliot, 8 May 1649, as printed in Henry Whitfield, *The Light Appearing More and More Towards the Perfect Day* (London, 1651).

[47] Thomas Thorowgood, *Jews in America* (London, 1660), The Learned Conjectures of the Reverend John Eliot Touching the Americans.

[48] Maclear, 'New England and the Fifth Monarchy', 246.

[49] This letter by Eliot was the longest statement by someone in North America on the origins of the aboriginals. For more, see Cogley, '"Some Other Kinde of Being and Condition"'.

millenarian' who developed a political model for the American aboriginals based on the ancient Israelite system, which he petitioned Cromwell to establish in England.[50]

The texts of Eliot, Thorowgood, Winslow and Menasseh fed into a growing apocalyptic environment in England that was fuelled by eschatological interpretations of the Thirty Years' War and the English civil wars,[51] as well as the end of effective censorship, which led to the reprinting of apocalyptic texts by English theologians such as Joseph Mede and Thomas Brightman.[52] All of which informed the perspectives of people, including John Dury, Edmund Hall and Nathaniel Homes, who argued that the discovery of the Israelites in the Americas proved that the end was upon them. Put together, this made these years the 'keenest and most widespread millennial expectancy' in England and America.[53]

Some of these individuals even gave specific dates to the anticipated end of the world. While Ralph Josselin, Nathaniel Homes, John Dury and Samuel Hartlib believed the final drama would occur in 1654 or 1655, the year 1656 was the most popular choice because it was thought to mirror the year of the flood.[54] As the Fifth Monarchist John Rogers stated, the flood had come in 1656 BC and lasted for 40 days; now fire would come to the world in 1656 AD and last for 40 years. Utilising John Archer's formulation, the Fifth Monarchist prophetess Mary Cary confidently claimed that 'in 1645 the Beast ceased to prevail against the Saints', the conversion of the Jews would happen in 1656 and Jesus' kingdom would come 'in its compleat glory' in 1701.[55]

Such beliefs led certain English Protestants to take an interest in contemporary Jews whose expected conversions made them crucial participants in the imminent onset of the millennium. By 1653, the Jews were an important topic of discussion,[56] which created a unique environment conducive to the campaign for Jewish

---

[50]   Richard Cogley, 'Idealism vs. Materialism in the Study of Puritan Missions to the Indians', *Method and Theory in the Study of Religion* Vol. 3, No. 2 (1991), 170.

[51]   Between 1642 and 1650, one 'cannot fail to be struck by the prevalance, among the leaders of the forces as well as among the rank and file, of the ideas that they were fighting the battles of Christ, and preparing for his kingdom'. See Louise Brown, *The Political Activities of the Baptists and Fifth Monarchy Men in England during the Interregnum* (New York: Burt Franklin, 1911), 14.

[52]   David Katz, *Philo-Semitism and the Readmission of the Jews to England 1603–1655* (Oxford: Clarendon Press, 1982), 93–4.

[53]   Richard Cogley, '"The Most Vile and Barbarous Nation of All the World": Giles Fletcher the Elder's "The Tarts or, Ten Tribes" (Ca. 1610)', *Renaissance Quarterly* Vol. 58, No. 3 (2005), 793; James de Jong, *As the Waters Cover the Sea: Millennial Expectations in the Rise of Anglo-American Missions 1640–1810* (Kampen: J.H. Kok, 1970), 37.

[54]   Todd Endelman, *The Jews of Britain, 1656 to 2000* (Berkeley: University of California Press, 2002), 20. For more on the importance of 1656 in English eschatological thought, see Christopher Hill, 'Till the Conversion of the Jews', in Richard Popkin, ed., *Millenarianism and Messianism in English Literature and Thought 1650–1800: Clark Library Lectures 1981–1982* (Leiden: E.J. Brill, 1988), 12–36.

[55]   Cary, *The Little Horns Doom and Downfall* (London, 1651), 202, 207. For more on Mary Cary, see David Loewenstein, 'Scriptural Exegesis, Female Prophecy, and Radical Politics in Mary Cary', *Studies in English Literature, 1500–1900* Vol. 46, No. 1 (2006), 141.

[56]   Katz, *Philo-Semitism and the Readmission of the Jews to England*, 189.

readmission to England. While Eliot's revelations concerning the Lost Tribes were one of the initial sparks that ultimately led to the return of the Jews in England,[57] it was the upsurge of interest in the Jews 'among the English Puritans and Fifth Monarchists more specifically' who thought the readmission was necessary for the conversion of the Jews that encouraged the campaign.[58]

The readmission campaign itself began in the Jewish–Christian circle in which Montezinos' testimony was first shared. Menasseh wrote, under Dury's influence, the *Humble Address* (1655) in order to petition Oliver Cromwell to allow the Jews to live in England. According to Menasseh, his serious consideration of the readmission was brought about by his Christian correspondents:

> Concerning the state of this my expedition, and negotiation at present, I shall onely say, and that briefly, that the communication and correspondence I have held, for some yeares since, with some eminent persons of England, was the first originall of my undertaking this design. For I always found by them, a great probability of obtaining what I now request; whilst they affirmed, that at this time the minds of men stood very well affected towards us; and that our entrance into this Island, would be very acceptable and well-pleasing unto them. And from this beginning sprang up in me a semblable affection, and desire of obtaining this purpose.[59]

Henry Jessey, a Fifth Monarchist preacher who anticipated the imminent conversion of the Jews and wrote the introductory remarks to Mary Cary's *The Little Horns Doom and Downfall*, was one of these 'eminent persons of England'. He played such a vital role in the readmission process that he has been labelled 'among the greatest friends of Israel in the early modern period'.[60]

News of the readmission campaign spread far and wide, reinforcing the messianic and millenarian tension that had originally promoted it. The Quaker missionary John Stubbs met Jews in the Italian peninsula who 'delight to hear of any hopes of an admission for them to live in England'.[61] Stubbs too framed the readmission in

57   Katz, *Philo-Semitism and the Readmission of the Jews to England*, 94, 84.

58   Christopher Hill, *Antichrist in Seventeenth-Century England* (Oxford: Oxford University Press, 1971), 107. Although this eschatological context was important in furthering the readmission campaign, Cromwell's pressing financial and commercial needs were also influential. For more, see Katz, *Philo-Semitism and the Readmission of the Jews to England*.

59   Menasseh ben Israel, *Vindicae Judaeorum or a Letter in Answer to Certain Questions Propounded by a Noble and Learned Gentleman* (London, 1656), 37.

60   David Katz, 'Philo-Semitism in the Radical Tradition: Henry Jessey, Morgan Llwyd, and Jacob Boehme', in Johannes van den Berg and E.G. van der Wall, eds, *Jewish–Christian Relations in the Seventeenth Century: Studies and Documents* (Dordrecht: Kluwer Academic Publishers, 1988), 195. For more, see David Katz, 'Henry Jessey and Conservative Millenarianism in Seventeenth-Century England and Holland', in Jozeph Michman, ed., *Dutch Jewish History: Proceedings of the Fourth Symposium on the History of the Jews in the Netherlands 7–10 December – Tel-Aviv-Jerusalem, 1986* (Jerusalem: 'Graf-Chen' Press, 1987), 75–93.

61   LSF Port MS 17.77: John Stubbs to his Quaker Friends in London, 29 March 1658.

terms of the hoped-for conversion of the Jews: he thought that it 'might tend much to the conversion of some among them if such a thing might come to pass'.[62] Even in the Americas, John Eliot understood the negotiation to allow the Jews to return to England in relation to the prophecy in Deuteronomy 28:64 in which God promised to scatter the Jews among the nations.[63] Like other Englishmen, he believed that this prophecy could not be fulfilled and the eschatological sequence could not proceed further until the scattering of the Jews included the British Isles.[64] Thus, the growth of interest in the Lost Tribes among a select group of Jews and Protestants on both sides of the Atlantic coupled with changing political conditions in England led to growth of eschatological tension that helped to promote the readmission campaign – a campaign which itself was then interpreted in a manner that spawned more millenarian and messianic excitement.

The Failure of the Millennium and its Long-term Effects

Like many English millenarians, John Eliot's belief in the Lost Tribes was tied to the much anticipated end of history. But the years 1654, 1655 and then the long-awaited 1656 came and went without the conversion of the Jews or Jesus' second coming. The continual disappointment had dire consequences. Eliot wrote to Thorowgood in the autumn of 1656 that he was withdrawing his endorsement of the Lost Tribes theory. He never would have taken such a strong stance, he said, if it had not been for Thorowgood's insistence.[65]

Despite the failure and disillusionment felt in these years, the bonds between the Jews and Protestant philo-Semites forged through these interactions had long-term consequences. Alongside increased correspondence and meetings between Menasseh and his Christian friends about the Lost Tribes and the readmission, these Protestants became acquainted with another rabbi: Nathan Shapira from Jerusalem. Shortly before departing for England, Menasseh met Shapira, who had been sent as an emissary to collect money for the Jews in the holy land from the Sephardic Diaspora. Menasseh even showed Shapira's letter to Cromwell as proof that the Jews needed a new place to live because they were being treated poorly in the Ottoman realms.

Meanwhile, Henry Jessey, Petrus Serrarius, John Dury and Samuel Hartlib all actively helped Shapira raise money. In Amsterdam, Serrarius engaged the rabbi

---

[62]   LSF Port MS 17.77: John Stubbs to his Quaker Friends in London, 29 March 1658.

[63]   See Thorowgood, *Jews in America*, The Learned Conjectures of the Reverend John Eliot Touching the Americans.

[64]   Eliot's focus on England was indicative of Englishmen who expected their country to play a special role in the coming end of history. The Fifth Monarchist pamphlet *A Door of Hope* insisted that the biblical prophecies must refer to England because even foreign writers recognised that the unparalleled upheavals in the country foreshadowed the approaching end. For more, see *A Door of Hope: Or, A Call and Declaration for the Gathering Together of the First Ripe Fruits unto the Standard of our Lord King Jesu* (London, 1660), 16.

[65]   Cogley, *John Eliot's Mission to the Indians before King Philip's War*, 96.

from Jerusalem in religious discussions and, upon learning of Shapira's views of the messiah, wrote: 'it seemed to me, that I did not hear a Jew, but a Christian … [who] was admitted to the inward mysteries of our Religion'.[66] Shapira may have been the Jew from Cracovia (Krakow) who convinced the Dutch chiliast that the redemption of Israel was drawing near;[67] a belief that would come to the fore a decade later when Serrarius became a vocal advocate for the restoration of the Jews to Palestine.

The interactions brought about by Shapira's mission to Amsterdam highlight multiple links between the Protestant philo-Semites and many of the key figures in the Sabbatian movement of the mid-1660s. Jessey was friends with Dury, Hartlib and Serrarius, who supplied the English Christians with much of the information about Sabbatai that reached Europe from the Ottoman Empire. Jessey was also in communication with Raphael Supino, a rabbi from Livorno who had accompanied Menasseh on his mission to London and would be an important source of Sabbatian news too. At the same time, the rabbi Jacob Hagiz, the teacher of Nathan of Gaza and excommunicator of Sabbatai, was aware of Jessey and his Puritan compatriots' work to raise money for the Jews.[68]

During these years, the same Christians fundraised for missions in the Americas. Alongside their frequent correspondences with the Puritan missionary, Jessey, Thorowgood and others sent Eliot £200-worth of merchandise between 1651 and 1657.[69] For the English millenarians, there was a direct connection between their work with the Jews in the Levant and their support for the Christianisation of the aboriginals. These distant places and different people were part of a single storyline in which the fulfilment of prophecy would come about across the world through their diligent efforts in England.

There were other significant ties between the messianism and millenarianism of this period and the Sabbatian movement of the following decade. In England, Thorowgood's *Jews in America* was reprinted as *Vindiciae Judaeorum* in 1666 'almost certainly to capitalize' on the reports of Sabbatai's gathering of the Jews and the Lost Tribes to return to Palestine.[70]

In the Dutch Republic, Menasseh was close to people who would become important Sabbatians, including Isaac Aboab de Fonseca and Abraham Israel Percira. The advent of Sabbatianism also brought about the re-emergence of interest in Menasseh's work: at the peak of Sabbatai's popularity, a Dutch translation entitled

---

[66]   John Dury, *An Information Concerning the Present State of the Jewish Nation in Europe and Judea* (London: R.W., 1658), 13.

[67]   David Katz, 'English Charity and Jewish Qualms: The Rescue of the Ashkenazi Community of Seventeenth-Century Jerusalem', in Ada Rapoport-Albert and Steven Zipperstein, eds, *Jewish History: Essays in Honour of Chimen Abramsky* (London: Peter Halban, 1988), 254.

[68]   For more on these connections, see Katz, 'English Charity and Jewish Qualms'.

[69]   Cogley, *John Eliot's Mission to the Indians before King Philip's War*, 213.

[70]   Richard Cogley, 'The Ancestry of the American Indians: Thomas Thorowgood's *Iewes in America* (1650) and *Jews in America* (1660)', *English Literary Renaissance* Vol. 35, No. 2 (2005), 324. This title was probably taken from Menasseh ben Israel's 1656 work of the same name.

De Hoop van Israel (1666) was printed twice in order to enlighten the Dutch public about Jewish messianism.[71]

In the Ottoman Empire, Menasseh's allure as a famous rabbi with a Converso background led other former Conversos to reprint the *Esperanca de Israel*. Originally owners of a printing house in Livorno, the Gabbai family set up a printing press in Smyrna in 1658.[72] It was there that the rabbi and printer Abraham Gabbai published the *Esperanca de Israel* after travelling to the Dutch Republic, where he may have personally acquired a copy of it himself.[73]

Spread among the Smyrnan Jewry, Menasseh's messianic message could have had an impact on the young Sabbatai; it is well known that the Conversos involved in the publishing of the *Esperanca de Israel* in Smyrna, such as Moses Pinheiro, were influenced by Menasseh's text, were friends of Sabbatai in his youth and would later become enthusiastic Sabbatians:[74] 'the large and important community of Portuguese Marranos living in seventeenth-century Smyrna, which, among other things, was unquestionably influenced by Ben Israel's *Esperanca de Israel*'.[75]

By Sabbatai's appearance in 1665, the *Hope of Israel* had been published in seven languages,[76] and 'it seems clear that there was a subtle but undeniable relationship between Christian and Jewish messianism, between these two and the intellectual atmosphere in the Portuguese Marrano communities in both Europe and the Orient – in particular, in Smyrna – and between both and the outbreak of Sabbatean messianism'.[77] Indeed, Menasseh's *Hope of Israel* points to the existence of 'a meeting point' between Christian and Jewish messianism,[78] and one 'might wonder whether Durie realised that his simple inquiry with Menasseh in 1649 had been one of the initial factors, by occasioning the publication of an influential book, of the messianic outburst of the 1660s'.[79]

The Impact of the Sabbatian Movement on John Eliot

Jews anticipated the return of the Lost Tribes of Israel alongside the coming of the messiah, so it should not be surprising that news of Sabbatai's messiahship was accompanied by rumours that the Lost Tribes had surfaced from their hidden

[71]   Mechoulan and Nahon, *Menasseh ben Israel*, 91.

[72]   Barnai, 'The Sabbatean Movement in Smyrna: The Social Background', 116.

[73]   Scholem, *Sabbatai Sevi*, 150; Barnai, 'Christian Messianism and the Portuguese Marranos', 121.

[74]   Goldish, *The Sabbatean Prophets*, 49; Barnai, 'Christian Messianism and the Portuguese Marranos', 120, 123.

[75]   Barnai, 'Christian Messianism and the Portuguese Marranos', 121. According to Barnai (122), 'The story of the ten lost tribes was disseminated (by way of Ben Israel's *Esperanca*) in Smyrna in 1659, where Sabbatai Zevi was then living' as well.

[76]   Barnai, 'Christian Messianism and the Portuguese Marranos', 120.

[77]   Barnai, 'Christian Messianism and the Portuguese Marranos', 123.

[78]   Barnai, 'Christian Messianism and the Portuguese Marranos', 120.

[79]   Van der Wall, 'Three Letters by Menasseh ben Israel to John Durie', 60.

dwelling, which were disseminated widely in printed pamphlets, gazettes and avvisi in Italian, Dutch and English, as well as moved across the Atlantic along Puritan channels in handwritten letters. Unlike the discovery of a single tribe located in the jungle in 1648, the stories in 1665 told of the appearance of mysterious armies of ancient Israelites in Persia, the Arabian Peninsula, North Africa and even the British Isles.[80]

When these rumours reached John Eliot in New England, he was enthralled. Although he had withdrawn his endorsement of the Lost Tribes theory, he thought that these were different: they were news that was confirmed from many places and printed in trusted sources. His earlier belief that the millennium was at hand had been right all along. It only needed to be modified from 1656 to 1666. Once again, Eliot became interested in the Lost Tribes.[81]

Like all of the previous accounts, these ones too would eventually be proven false. Yet Eliot would not be entirely dissuaded this time. He remained intrigued by the Lost Tribes for the rest of his life.[82]

Eliot's evolving beliefs about the Lost Tribes were a product of a broad historical environment that included Jews and Christians across the Atlantic and Mediterranean worlds. Eliot was a local actor, living in North America, who was influenced by far-reaching intellectual currents.

Like the Christian millenarianism in which he believed, Eliot himself is often located within a transatlantic English framework. The English context was important: the parliament created a committee to gather and promote stories about New England missionary activity among the aboriginals in 1649, Menasseh's *Hope of Israel* was dedicated to the parliament in 1650, and Eliot dedicated his 1653 pamphlet on the conversion of indigenous peoples to Cromwell. Yet it is only a part of the story.

One cannot make sense of Eliot's changing views without examining the multiple threads of narratives that were woven together in different patterns as they were moved among and between networks of Jews, Judaising Conversos, Presbyterians and Puritans from one side of the Atlantic to the far side of the Mediterranean and back again. Eliot's original belief that the Israelites were in Asia was influenced by a history of Puritan eschatological thought and dialogue that connected England and New England. The argument that he repeated, outlined by his Boston pastor John Cotton, originated in Europe and was advocated by English Protestants such as Thomas Brightman and John Archer.

Eliot changed his mind and became a proponent of the theory that the aboriginals were the descendants of the ancient Hebrews largely because of a chain of events to which he was only remotely linked: events that included the journey of a Converso from South America to Amsterdam, the Converso's dialogue with a Dutch rabbi and the rabbi's discussion with an itinerant Scotsman who then met a Presbyterian minister in England who had, remarkably, been influenced by Eliot's writings himself.

[80] For more on this topic, see Chapter 3, 'Who Sacked Mecca? The Life of a Rumour (1665–1666)'.
[81] Cogley, 'John Eliot and the Origins of the American Indians', 221–2.
[82] Katz, *Philo-Semitism and the Readmission of the Jews to England*, 157.

The numerous interactions between the Englishmen and Dutch rabbi produced publications that led the Puritan missionary to consider the apocalyptic consequences of his own work. They also had ramifications in the political and religious arenas. In the former, they encouraged the Jewish readmission to England. In the latter, they fed into an environment that promoted the intensification of Christian millenarianism and Jewish messianism.

Although the long-awaited millennium did not arrive in 1656, disheartening Eliot and prompting him to abandon his advocacy of the Lost Tribes theory, Menasseh's *Hope of Israel* may have played a role in the Sabbatian movement. At the very least, the Jewish–Christian connections forged in the late 1640s and early 1650s furthered cross-religious exchanges that would be utilised in the transmission of Sabbatian information in 1665 and 1666.

It was the rumours of the Lost Tribes that were spread alongside the news of Sabbatai in the mid-1660s that caused Eliot to reconsider his ideas again. Quite simply, the reports reinvigorated his earlier beliefs. This chapter has only briefly touched upon this final process because the history of the rumours of the Lost Tribes in 1665 comprises the entire third chapter of this text. Before we discuss the movement of these accounts from the Middle East across Europe to the Americas, it is necessary to examine the transnational transmission of an intermediary eschatological construct: news of the Quaker leader James Nayler's messianic entrance into Bristol in 1656.

Chapter 2

New Monarchs or Grand Impostors?
James Nayler and Sabbatai Sevi
(1656–1666)

Here stands James Nayler as he should be dressed,
The Quakers' chosen monarch in the west:
And here is Sabatai, who in the east
The Jews' Messiah reigns, their King and Priest.[1]

These lines, inscribed on a woodcut in *Anabaptisticum et Enthusiasticum Pantheon und Geistliches Rust-Hauss* (1702), deride the two 'Monarchia Nova' the English Quaker James Nayler and the Ottoman Jew Sabbatai Sevi. Although shown standing side by side, Nayler and Sevi neither met nor travelled to any of the same places. Indeed, Sevi was not known outside of the Jewish world until 1665 whereas Nayler, who died in 1660, was principally known in England for his actions committed almost a decade earlier.

It was in the autumn of 1656, the year many English millenarians expected Jesus to return, that Nayler and his small band of followers approached the city gates of Bristol.[2] Nayler sat on horseback as Martha Simmonds, Hannah Stranger, Dorcas Erbury and a few others trudged through the mud alongside him, crying 'Hosanna, Hosanna, Hosanna in the highest'.

In seventeenth-century England, what they were doing was considered scandalous. News that they were replicating Jesus' entry into Jerusalem spread quickly throughout the city, and the Quakers were brought before the magistrates of Bristol to explain themselves. In front of the judges, Nayler's followers showed their audacity. Martha Simmonds and Hannah Stranger told them that Nayler was Jesus while Dorcas Erbury claimed that Nayler had raised her from the dead.

The magistrates were not impressed. Nayler and most of his followers were escorted under guard to London, where the Quaker leader was questioned before the

---

[1] The translation of this quotation from *Anabaptisticum et Enthusiasticum Pantheon und Geistliches Rust-Hauss* (Hamburg, 1702) is found in Mabel Richmond Brailsford, *A Quaker from Cromwell's Army: James Nayler* (London: The Swathmore Press Ltd., 1927), 187.

[2] According to Richard Popkin, *The Third Force in Seventeenth Century Thought* (Leiden: E.J. Brill, 1992), 113, the Quakers also established a mission in Amsterdam in 1656 because of the millenarian predictions regarding the conversion of the Jews in this year.

Figure 2.1   Woodcut of James Nayler and Sabbatai Sevi in *Anabaptisticum et Enthusiasticum Pantheon und Geistliches Rust-Hauss* (1702). Courtesy of Bibliotheca Rosenthaliana, Special Collections of the University of Amsterdam.

English parliament. Throughout his examination, Nayler admitted the facts but stated that his act only had allegorical significance.[3] After much deliberation, a verdict was delivered: Nayler was guilty of blasphemy. A hole was bored in his tongue, the letter 'B' for blasphemer was branded on his forehead and, after being whipped through the streets of London and Bristol, he was imprisoned. News of Nayler's actions and punishment was disseminated far and wide. From the American colonies to Poland, people of a variety of religious and national backgrounds knew about James Nayler.

Ten years later, word of Sabbatai Sevi's messiahship reached the same Christian populations, and some of these people saw connections between the actions of the English Quaker and the Ottoman Jew. Alongside the woodcut of the two men in *Anabaptisticum et Enthusiasticum Pantheon und Geistliches Rust-Hauss*, John Evelyn's *History of the Three Late Famous Impostors* (1669) presented Nayler's biography next to Sevi's, a German pamphlet titled *Beschreibung Des Newen Judischen Konigs Sabetha Sebi* (1666) labelled Sevi 'a Turkish [Moslem] or a Jewish Quaker', and a Polish correspondent from Amsterdam wrote of a boatload of Quakers who left Bristol for Jerusalem simply to find out more about the Sabbatian movement.[4]

Sources such as these have led historians to make claims about the potential cross-religious impact of Nayler's messianic actions on Sevi and the outbreak of Sabbatianism. Richard Popkin has argued that Sabbatai 'may have been influenced by the earlier Quaker Messianic claimant. His father worked for Quaker merchants'.[5] He has also written that,

> David Katz and I have been following possible connection between the Nayler movement and the prelude to Sabbatai Sevi. There were Quaker merchants in Smyrna, and Sabbatai's father is supposed to have worked for one of them. Possibly more interesting is that a couple of Quaker missionaries went off to convert the Pope and the Sultan ... In a report it is said that they were arrested for preaching the imminent coming of the Messiah. Their arrest occurred when Sabbatai was in Jerusalem.[6]

In spite of such statements, there has yet to be a sustained investigation of the possible ways in which Nayler and the Quakers could have affected the outbreak of the Sabbatian movement. This chapter examines the networks along which narratives of Nayler's entry into Bristol and the messianism associated with the Quaker leader were disseminated across Europe and beyond. Doing so, on one hand, proves that news of

---

[3]   Ivan Roots, *The Great Rebellion 1642–1660* (London: B.T. Batsford Ltd., 1972), 206.

[4]   Heyd, 'The "Jewish Quaker"', 234; Scholem, *Sabbatai Sevi*, 548.

[5]   Richard Popkin, 'Seventeenth-Century Millenarianism', in Malcolm Bull, ed., *Apocalypse Theory and the Ends of the World* (Oxford: Blackwell, 1995), 122.

[6]   Popkin, 'Christian Interest and Concerns about Sabbatai Zevi', 94. Matt Goldish (202) has correctly pointed out that Popkin was mistaken about Sevi being in Jerusalem at this time; however, he added that the presence of Nathan of Gaza in Jerusalem is probably more interesting anyway, leaving the possibility of transmission and influence open to question. Other historians have touched upon this issue, if only in passing. A few of the 20 biographers of Nayler have dedicated a paragraph or two to this topic.

Nayler's actions travelled across Europe to an extent not previously recognised. On the other hand, it also shows that it is very unlikely that word of Nayler reached Sevi or Nathan and influenced the Sabbatian movement.

Chronicling the transmission of news related to this parallel but disconnected messianic movement highlights two asymmetrical relationships across the seventeenth-century Abrahamic world. First, news from the Ottoman Empire was spread widely across Europe, but the failure of the Nayler narrative to reach the Ottoman lands demonstrates that the reverse was not true. Secondly, the fact that the messianism associated with Nayler did not have an impact on the Jews points to the continuation of the dominant historical trend in which events in the Jewish world have long affected Christians whereas Christian influence on Jewish thought has been less profound. After all, Judaism is part of the Christian heritage, not vice versa.

English Broadsheets and Pamphlets

The story of Nayler was disseminated rapidly across Europe. In England, printed gazettes reported on Nayler's messianic entrance within a week, introducing much of the English population to the Quaker leader. Between late October and January, the two primary English news sources of the day, the *Mercurius Politicus* and the *Publick Intelligencer*, each printed at least 15 stories about Nayler. Sometimes there were numerous items in a single issue and, on more than one occasion, at such a length that they took up multiple pages. In their articles, the editors of the gazettes consistently described Nayler and his group negatively. Nayler was 'a grand Impostor, and a great Seducer of the people'; he was 'guilty of horrid Blasphemy'.[7]

In the following two months, the Quaker messiah was the subject of multiple full-length pamphlets that utilised the Nayler affair as an entry point for a broader attack on the Quakers. Ralph Farmer, a preacher who had lost to Nayler in debates, titled his treatise *Satan Inthron'd in his Chair of Pestilence* (1656) and presented Nayler as an 'Impostor' who had 'come this way to play [his] pranks with us'.[8] For Farmer, Nayler and his entourage were clearly evil: 'There have (and there may again) come false Christs, and Antichrists accompanied with the workings of Satan'.[9]

John Deacon's *The Grand Impostor Examined* (1656) came out shortly after Farmer's pamphlet, and it was neither as in-depth as Farmer's nor was it intended to be. Like Farmer, Deacon had nothing but sheer disdain for Nayler. He was 'a deluded and deluding Quaker and Impostor', a man 'of so erroneous and unsanctified a disposition, that it is hard to say whether heresie or impudencie beareth the greater rule in him'.[10]

These two pamphlets did not go unanswered. Shortly after their release, George Bishop, the well-known defender of the Quakers, published *The Throne of Truth*

---

[7]   *Mercurius Politicus,* 4 December 1656.
[8]   Ralph Farmer, *Sathan Inthron'd in his Chair of Pestilence* (London, 1656), 3.
[9]   Farmer, *Sathan Inthron'd in his Chair of Pestilence*, 41.
[10]   John Deacon, *The Grand Impostor Examined* (London, 1656), 40.

Exalted over the Povvers of Darkness (1657). Not concerned with Deacon's work, Bishop took apart Farmer's manuscript, discrediting it to defend the Quakers. He accused Farmer of 'the foul Forgery and dishonest dealing in such a matter of weight as his life, and the name of the Truth of the living God'.[11] Yet in order to protect the Quaker movement, Bishop turned his back on Nayler. Nayler had 'walked in the Light, and in it ruled' until 'his hour of Temptation being come, and Darkness getting about him quick and sudden'.[12] Such polemical sources spread Nayler's story to a broader English population and kept it in the spotlight long after it had ceased to be newsworthy.

Italian Diplomatic Reports about Nayler

News of Nayler crossed the English Channel thanks in part to the growth of the modern diplomatic system. At the time that the gazettes and pamphlets about Nayler were widely available to the English public, there were three Italian diplomats stationed in London who often turned to such sources to either confirm the reports they were hearing or as an entry point to the news.

Although the Italian diplomats were all doubtlessly aware of Nayler's actions and interview before parliament, they thought that the Quaker leader was worthy of different amounts of attention, as reflected in the time they took to inform their home governments of the affair. The Venetian diplomatic representative in London, Francesco Giavarina, cared little about Nayler and only indirectly referred to him in one of his regular dispatches to the Venetian senate:

> Thus it would seem that at present there is not a thoroughly good understanding between the Protector and the Parliament, and his Highness has taken occasion to display his vexation at some death sentences made by the Assembly without his consent. He has written a sharp letter to parliament in which, while admitting the sentences to be just, he blames the members for coming to these without seeking his advice and approval, which he declares are necessary since he is associated with them in the government of the State and in all things.[13]

Giavarina must have been confused because Nayler was not sentenced to death. Regardless, the Quaker leader was of no interest to the Venetian diplomat.

Unlike Giavarina, the Genoese resident Francesco Bernardi wrote specifically about Nayler in his diplomatic reports. While Nayler's entrance into Bristol occurred

---

[11]   George Bishop, *The Throne of Truth Exalted over the Povvers of Darkness* (London, 1657), 25.

[12]   Bishop, *The Throne of Truth Exalted over the Povvers of Darkness*, 3, 4.

[13]   According to the editor of the state papers, this letter undoubtedly referred to Nayler. Francesco Giavarina to the Venetian doge and senate, 12 January 1657, as translated and summarised in Allen Hinds, ed., *Calendar of State Papers Relating to English Affairs Existing in the Archives and Collections of Venice and in Other Libraries of Northern Italy* Vol. 31 (London, 1930): 1657–1659, 1–11.

in late October and English gazettes and pamphlets had appeared by December, Bernardi did not write about Nayler until 17 December. When he did, Bernardi described Nayler as a political rebel.[14] No longer viewed as the initiator of a 'prank', Nayler was part of a 'conspiracy' against the government: this 'so-called Christ' aimed for nothing less than the 'totale distruttione della Christianita et ogni buon governo' (total destruction of Christianity and every good government).[15] The Quakers had recently reached such a high degree of blasphemy that they raised 'un Capo, che chiamasi il secondo Christo' (a leader whom they called the second Christ) who walked diverse provinces with a great following and 'Maria e Magdalena' (Mary and Magdalene), committing the actions of 'nostro Redentore' (our redeemer) written about in the sacred scriptures.[16]

Being a diplomat, Bernardi focused on the political arena of which he was a part; his reports included discussions of the religious scene in England that were analysed from a strictly political view. Therefore, Bernardi defined Nayler explicitly as 'Un Masaniello Quacchero'.[17] Furthermore, Bernardi provided a view of Cromwell's interaction with Nayler that was unavailable in the gazettes. Writing in a manner that suggested he was at Nayler's interview before Cromwell himself, Bernardi quoted the protector turning to the Quaker leader and saying, 'Tu sei un seduttore del popolo' (You are a seducer of the people).[18]

Throughout, Bernardi employed the same terms and phrases as the English press. Nayler was guilty of 'horribil blasphemia' (horrible blasphemy) and of being 'un grandissimo impostore, et seduttore del popolo' (a grand impostor, and a seducer of the people).[19] Yet the Genoese diplomat claimed, on two occasions, that part of Nayler's punishment was to have his ears cut off, so he must not have drawn his information from the English gazettes.[20] Bernardi may have had better political access than the English news editors but he still inaccurately reported the basics of Nayler's punishment, which were widely known through these sources. While somewhat

---

[14]   Francesco Bernardi to the Republic of Genoa, 17 December 1656, as quoted in Stefano Villani, 'Un Masaniello Quacchero: James Nayler', *Rivista di Storia e Letteratura Religiosa* (Florence: Leo. S. Olschki, 1997), 77–8. For more on Franceso Bernardi, see this article.

[15]   Francesco Bernardi to the Republic of Genoa, 17 December 1656, as quoted in Villani, 'Un Masaniello Quacchero', 77–8.

[16]   Francesco Bernardi to the Republic of Genoa, 17 December 1656, as quoted in Villani, 'Un Masaniello Quacchero', 77–8.

[17]   Stefano Villani, 'Seventeenth-Century Italy and English Radical Movements', in Ariel Hessayon and David Finnegan, eds, *Varieties of Seventeenth- and Early Eighteenth-Century English Radicalism in Context* (Farnham: Ashgate, 2011), 158.

[18]   Francesco Bernardi to the Republic of Genoa, 17 December 1656, as quoted in Villani, 'Un Masaniello Quacchero', 77–8.

[19]   Francesco Bernardi to the Republic of Genoa, 17 December 1656, as quoted in Villani, 'Un Masaniello Quacchero', 77–8.

[20]   Francesco Bernardi to the Republic of Genoa, 17 December 1656, as quoted in Villani, 'Un Masaniello Quacchero', 77–8.

unexpected, a mistake such as this could have occurred – the removal of one's ears was common punishment.[21]

The third Italian diplomat in London was the most experienced. The Tuscan representative Amerigo Salvetti had lived in England for over 20 years and, after learning of the Bristol affair, wrote extensively about Nayler in six of his reports in December and January.[22] Although Salvetti stated that Nayler committed 'nefando biasteme' (nefarious blasphemy),[23] he did not use the common terms of 'Impostor' and 'Seducer' like most other English and Italian authors. Instead, Salvetti spoke of 'sua pazzia' (his madness) and his 'humori malinconici' (melancholy humours),[24] descriptions of Nayler which had no precedent.

Salvetti may have moved in the same political circles as Bernardi, but he was not concerned with Cromwell's views. For Salvetti, Nayler was important due to the political influence of the Quakers on the one hand and as a distraction to parliamentary affairs on the other. His commentary on Nayler immediately preceded comments about the future of the government: it would have to be negotiated delicately, especially with the opposition from the Presbyterians and Quakers. Salvetti also noted that the new messiah, on this occasion and others, was a cause of distraction between the religious sects.[25]

Italian Avvisi

All of the Italian diplomats sent their dispatches to their respective states, where news industries were starting to develop. Avvisi editors were renowned for acquiring much of their information from diplomatic dispatches and, as early as January 1657, these editors printed items about Nayler. The first Italian report on Nayler was found in a printed avviso from Genoa, which was reprinted in Florence. An early January issue stated that Cromwell 'aveva fatto carcerare 4 donne, & un' huomo, che lo chiamavano il Messia, qual parlava contro il presente governo' (had imprisoned four women and a man, whom they called the Messiah and had spoken against the present government).[26] Although these nameless people can be none other than Nayler and his circle, the origins of the account are hard to identify. The avviso claimed to draw

---

[21] John Traske, for example, had the letter 'J' for Judaiser branded on his forehead and his ears nailed to the pillory.

[22] For more on Amerigo Salvetti, including full copies of his diplomatic correspondence, see Stefano Villani, ed., *La Corrispondenza dei Residenti Toscani a Londra: Commonwealth e Protettorato (11 Dicembre 1648 – 11 Giugno 1660)*. Unpublished manuscript.

[23] Amerigo Salvetti to Giambattista Gondi, 26 January 1656, as quoted in Villani, *La Corrispondenza dei Residenti Toscani a Londra*.

[24] Amerigo Salvetti to Giambattista Gondi, 12 January 1656, as quoted in Villani, *La Corrispondenza dei Residenti Toscani a Londra*.

[25] Amerigo Salvetti to Giambattista Gondi, 12 January 1656, as quoted in Villani, *La Corrispondenza dei Residenti Toscani a Londra*.

[26] SA Segreteria di Stato, Avvisi, 27, 10: Genoa, 6 January 1657.

its information from letters from London, but these were most likely not from the diplomats, who only sent their correspondence after 17 December and who provided names and details.

A month later, another printed article about Nayler surfaced in the Italian press, this time in Milan:

> The so-called head of the Quakers (one of the new sects in that kingdom) called James Naylor, has been convicted by the parliament in London on the 22nd of December as a seducer of the people because he claimed to be the messiah. He is to be brought through the major areas of the city to that infamous place where, having a hot iron passed through his tongue and put on his forehead as a blasphemer, he is to be sent to Bristol to do the same and then back to London to be imprisoned for life. Daily one can hear both the news and the unpleasant effects of the religions that are all harmful and contrary to the true Roman Catholicism as well as to the design of Cromwell himself.[27]

This news item was much more detailed. It provided Nayler's name, labelled him the head of the Quakers and told of his punishment for divulging himself as the messiah and for being a seducer of the people. The avviso editor also placed the story in an Italian religious context: Nayler's actions were harmful and contrary to both Cromwell's designs and the true Roman Catholic faith.[28]

These details suggest that, like the earlier avviso from Genoa, this one was not based on any of the diplomatic dispatches. The accurate description of Nayler's punishment meant that it did not come from Bernardi while the usage of 'seducer of the people' meant that it did not come from Salvetti either. Instead, it most closely resembled the reports found in the English gazettes, which could have been reprinted in Spanish or French gazettes that made their way to Italy.

Although items about Nayler were published in printed avvisi in multiple Italian states with diplomatic representatives in England, there were no connections between the accounts written in the English gazettes, the Italian diplomatic correspondence and the Italian avvisi.[29] The transmission of the Nayler narrative therefore highlights disjunctures in the movement of information between England and Italy, even among sources that would be expected to be linked.

---

[27]   BNCF Codd Magliabechiani XXV, 738, 74b: Milan, 7 February 1657. The original stated: 'L'avvisato Capo delli Tremboli (una delle Sette nuovamente suscitate in quell Regno) nominato Giacomo Naylor, in pena d'essersi divulgato per il Messia, era stato alli 22. di Decembre condannato dal Parlamento di Londra, come sedutore del Popolo, ad esser condotto in public in un luogo infame, fustigato per le contrade maggiori di quella Città, trapassatagila lingua con un ferro infuocato, e col medesimo marcato in fronte, come biastermatore, per dover esser poscia mandato à Bristol sua Patria à far il medesimo, & d'indi di nuovo à Londra in Carcere perpetua, mà ciò non ostante giornalmente si sentivano nuovi, e grand'inconvenienti di Religione, per esser tutte dannose, e contrarie alla vera Catolica Romana, non men che a disegni dello stesso Cromwell'.

[28]   BNCF Codd Magliabechiani XXV, 738, 74b: Milan, 7 February 1657.

[29]   There are no extant Venetian avvisi that mention Nayler, but that could be because very few from this period survive.

Latin Texts

Alongside the avvisi, Latin texts about Nayler circulated around Italy. In 1659, the Irish Franciscan Maurice Conry published *De Extremis Anglo-Haereticorum*, which presented the history of the many sects of English 'heretics'. In the course of his discussion on the Quakers, Conry dedicated a couple of lines to Nayler: 'Habent quondam, haeresiarcham Jacobum Nayler, quem nonnulli ipsorum, Christum, Salvatorem mundi, et Jesum appellarunt, et pro tali coluerunt' (there was a certain heretic named James Nayler, who people had called Jesus Christ, the Saviour of the World, and worshipped him as such).[30]

A decade later, the Franciscan Anthony Bruodin published *Propugnaculum Catholicae Veritatis* (1669). In a subsection of this text, entitled 'De Quakerorum Secta', Bruodin chose Nayler as an example of the madness that occurs with the abandonment of the Catholic Church. Nayler, who had the support of his followers and the admiration of all, was 'a quo circa fundamento malignae suae sectae' (of the fundamental wickedness of his sect).[31] He was a 'Smigmator' (a concocter or rabble-rouser), a 'Hispidus' (a word mainly used to describe animals as hairy and brutish) and an 'Insatiabilis Helluo' (insatiable glutton) who was devoted to carnal pleasures and renowned for a thousand other terrible transgressions.[32] By selling himself as Jesus with his voice, appearance and gestures, Nayler had garnered the devotion of his disciples, who worshipped him as the saviour of the world. All of this happened, according to Bruodin, through the powers of the antichrist.[33]

Despite being printed 10 years apart, most of the information about Nayler in both texts came from Conry, who had been imprisoned in Bristol and London until 1658. By chance, Conry was incarcerated alongside Nayler, as he later recounted: 'vidi ipsum in caceribus Bristoliae, et saepissime locutus sum illi' (I have seen him in prison at Bristol, and with wisdom I have spoken to him).[34] After serving as a fellow prisoner of Nayler, Conry returned to Rome, where he participated in the trial, most likely as a translator, of the two Quaker missionaries John Perrot and John Luffe before the Inquisition.[35] Not only did Conry take part in the religious persecution of the

---

[30]   Maurice Conry, *De Extremis Anglo-Haereticorum* (1659) as printed in Stefano Villani, 'Appendix: Documents Relating to the Hat Controversy', *Benjamin Furly 1646–1714: A Quaker Merchant and His Milieu* (Florence: Leo. S. Olschki Editore, 2007), 185–6.

[31]   Anthony Bruodin, *Propugnaculum Catholicae Veritatis* (1669) as printed in Villani, 'Appendix: Documents Relating to the Hat Controversy', 192.

[32]   Bruodin, *Propugnaculum Catholicae Veritatis* as printed in Villani, 'Appendix: Documents Relating to the Hat Controversy', 192.

[33]   Bruodin, *Propugnaculum Catholicae Veritatis* as printed in Villani, 'Appendix: Documents Relating to the Hat Controversy', 192.

[34]   Conry, *De Extremis Anglo-Haereticorum* as printed in Villani, 'Appendix: Documents Relating to the Hat Controversy', 185–6.

[35]   Stefano Villani, 'Conscience and Convention: The Young Furly and the Hat Controversy', in Sarah Hutton, ed., *Benjamin Furly 1646–1714: A Quaker Merchant and His Milieu* (Florence: Leo. S. Olschki Editore, 2007), 107–8.

very same group that he suffered with in England, but he also shared his experience with his Franciscan brother Bruodin, who later used it in his own polemic against the Quakers.[36]

Dutch Newsbooks and Poems

Like in the Italian peninsula, multiple types of documents in the Dutch Republic published accounts of Nayler's messianic entrance into Bristol. Due to the Dutch press' close proximity to England and its complex relationship with the English gazettes, the Dutch newsbook titled the *Hollandtze Mercurius* was one of the first European sources to print Nayler's story. As early as December, the Dutch newsbook told the well-known history of the Quaker leader, describing him as an 'Impostor' and providing his background: Nayler was a Quartermaster under Major General Lambert in Scotland and a soldier under Fairfax – supplementary information found in none of the Italian diplomatic dispatches or avvisi.[37] The *Hollandtze Mercurius* also had a broader geographical perspective than most other sources: it turned directly from its discussion of Nayler to seven Quakers who were recently discovered in Boston.[38] Nayler was not a big deal himself, but he was indicative of a larger movement that was spreading eastward across Europe and westward across the Atlantic Ocean.

Alongside the newsbook, the story of Nayler was published in the Dutch pamphlet literature. It became so 'widespread in Holland and brought the Quakers there much notoriety and difficulty' because Nayler's claims were used 'to prove that the Quaker's heresy was blasphemous in the extreme'.[39] In fact, Quakers in the Dutch Republic were forced to defend themselves against the charges associated with Nayler for years after, which were published so frequently in Dutch that the missionary William Ames had to economise time and space in responding to them. In his *Antichrist Uncovered*, for example, he replied to three different Dutch attacks on Nayler, one of which was the Dutch translation of John Deacon's *The Grand Impostor Examined* that came out in 1657.[40] Others were based on independent research, such as Jacob Adriaensz' pamphlet entitled *Extracts (or Abstracts) from the writings of Jacob Adriaensz. Serving as instruction for some who permit themselves to be called Quakers (or Shakers)*, which was based on an interview that Adriaensz had with Nayler whom he visited in prison in England.[41]

Not all of the non-English publications portrayed Nayler negatively. While the Quaker William Caton wrote to Margaret Fell about a 19-page book in the Dutch

---

[36]   Villani, 'Conscience and Convention', 107–8.

[37]   ULL Castelyn, *Hollandtze Mercurius*, 126: December 1656.

[38]   ULL Castelyn, *Hollandtze Mercurius*, 126: December 1656.

[39]   William Hull, *The Rise of Quakerism in Amsterdam, 1655–65* (Philadelphia: Patterson and White Company, 1938), 230.

[40]   Hull, *The Rise of Quakerism in Amsterdam*, 237, 244, 241.

[41]   Hull, *The Rise of Quakerism in Amsterdam*, 237.

Republic that attacked the Quakers by giving the relation of Nayler's trial and the transcription of letters from his adherents that referred to him as Jesus,[42] a Dutch pamphlet entitled *Klachte der Quakers, Over Haren Niewen Martelaer, James Nailor in Engelandt* (1657) used verse 'to exploit Nayler's story to the utmost and to portray the Quaker's grief'.[43] While it is not known whether this poem was composed by a Dutch writer or whether it was based upon an unknown English source,[44] the frontispiece bears a portrait of a somewhat stern and dejected 'Nailor'.

The pamphlet contained a 68-line poem that described Nayler as the Quaker's 'Nieuwen Martelaer' (New Martyr), 'groote Sant' (great Saint) and 'vrome Quakerbend' (devout Quaker).[45] His ride into Bristol had made their spirits and feet dance. Although the branding of the letter 'B' on Nayler's forehead was supposed to mark him permanently as a blasphemer, this poem reframed the punishment: 'Zijn brantmerk strekke een heylig teeken' (He was branded with a holy sign).[46]

While a number of the English and Dutch pamphlets were spread to the German lands where they were published in German translations,[47] Nayler's story was also known of in France; an English traveller in Paris wrote to Joseph Williamson in London:

> We have now arrived at Paris, after being much cheated by the zealous reformists, under cloak of religion, especially at Montauban. I suppose you have the news of one Naylor, a Quaker in England, who pretended to be the Messiah, and carried about with him 12 apostles, and 2 sinful Magdalens; but that fancy will be jerked out of him by his sentence. I wish his apostles the same persecution. Two of that set lately passed by Paris, and were found starving in the streets, but some English gentleman, not knowing that religion, relieved them. They said they were ambassadors from the Lord to the Duke of Savoy, and despaired not of the gift of tongues, for the Lord had told them they should have success.[48]

While referencing Nayler's female followers as 'Magdalens' was not unprecedented, this letter told of Nayler's 12 apostles – figures not mentioned anywhere else who could be fictitious characters added to the story. After all, Nayler was seen to be mimicking Jesus so such details, which are found in the various versions of Jesus' life in the Bible, could have been appended to the account of Nayler during circulation. In other words, the seventeenth-century news was possibly being informed by the biblical account.

---

[42]   Brailsford, *A Quaker from Cromwell's Army*, 186: see LSF Swarthmore MS IV, 370. This may be *The Devil changed into a Quaker* that, according to Hull, *The Rise of Quakerism in Amsterdam*, 190, was a Dutch version of an English account of James Nayler's 'blasphemies' published, possibly in Utrecht, in 1657.

[43]   Hull, *The Rise of Quakerism in Amsterdam*, 247.

[44]   Hull, *The Rise of Quakerism in Amsterdam*, 249.

[45]   ULL *Klachte der Quakers, Over Haren Nieuwen Marterlaer, James Nailor in Engelandt* (1657), 2.

[46]   ULL *Klachte der Quakers, Over Haren Nieuwen Marterlaer, James Nailor in Engelandt* (1657), 4.

[47]   Hull, *The Rise of Quakerism in Amsterdam*, 255.

[48]   Chas. Perrott to Joseph Williamson, 7 January 1657, as quoted in *Calendar of State Papers Domestic: Interregnum* Vol. 153: 1656–7, 223–58.

To the Ottoman Empire?

James Nayler was not the only messianic figure that people thought Joseph Williamson, the English undersecretary of state, would be interested in. A decade later, Williamson received numerous letters from English merchants and diplomats in the Ottoman Empire and the Italian peninsula about the Jewish messiah Sabbatai Sevi.

Sevi was a young man in the Levant when Nayler entered Bristol. Unlike in Europe, neither gazettes nor pamphlets were prevalent in the Empire so there were no news sources in which Sevi could have read about Nayler. Moreover, the Ottoman authorities did not have any representatives in England, so neither was there an Ottoman diplomatic correspondence that transmitted the narrative across the Mediterranean.[49] Finally, judging by their extant letters, the Europeans stationed in the Empire do not appear to have been informed of the messianic entrance of Nayler in these years. As we shall see, even though many knew about the Quakers, they did not mention Nayler. Thus, no known evidence proves that the story of Nayler reached Sevi or Nathan in the Levant, in spite of the aforementioned remarks of certain historians.

There were, however, numerous networks that connected the European countries to the Ottoman Empire and, because the seventeenth-century historical record is incomplete and scholars have suggested that transmission and influence occurred, the remainder of this chapter examines these networks in order to show that it was unlikely the messianism associated with James Nayler was brought to the Levant and, even if it was, it is highly improbable that it affected Sabbatai Sevi, Nathan of Gaza or the outbreak of the Sabbatian movement.

Nayler's Followers

It has been alleged that Nayler's 'unrepentant followers fled to Holland and to the outposts of Quaker trading in the Levant, and in the western hemisphere, and they seemed to have carried their millenarian message with them.'[50] If the Quaker

---

[49]   Secret Ottoman correspondence may have moved the story through government channels. 'A Turkish View of Quakerism, 1659', *Journal of the Friends Historical Society* Vol. 8 (1911), 25–7 discusses and presents the account of the fictitious *Letters Writ by a Turkish Spy who Lived Five and Forty Years Undiscovered in Paris, 1637–1682.* This text, which Popkin, *Isaac La Peyrere*, 124, stated came from an Ottoman spy in Paris who reported the Nayler episode to Istanbul, claims that 'the ringleader of this people professes himself to be the Messias, being, in all parts of his body and features of his face, like Jesus the son of Mary ... If thou wouldest have my opinion on these new religionists in Europe, and their leader; I take him to be an impostor, and his followers to be either fools or madmen ... they are not men of arms, but a herd of silly, insignificant people, aiming rather to heap up riches in obscurity, than to acquire fame by an heroick undertaking. They are generally merchants or mechanicks.' While the notation that the followers are merchants is noteworthy, even if this report made it to Istanbul, it presented Nayler negatively and would have been hard pressed to reach Sevi.

[50]   Popkin, 'Seventeenth-Century Millenarianism', 120. In a work published a decade earlier, Popkin tempered his claims by stating that Nayler's followers sent missionaries as far as France and New England to report that the messiah had arrived. Richard Popkin, 'Spinoza, the Quakers and the Millenarians, 1656–1658', *Separata de Manuscrito* Vol. 6, No. 1 (1982), 123.

messianism came to the Ottoman lands with enough force to convince people of its veracity, the primary carriers would have been those involved in Nayler's messianic entrance into Bristol. Yet there is no evidence of any of Nayler's followers travelling to the Levant. Nayler, upon his release in 1659, settled into the home of Rebecca Travers in London and died in the north of England in 1660.[51] Nayler's leading follower Martha Simmonds, the '*fons et origo mali*' of the messianic drama at Bristol,[52] could have been one of these 'unrepentant followers' except she became quiet after reading the paper that Nayler produced in 1657, which condemned the spirit of the disorder in 1656.[53] Giving up her messianic beliefs, Simmonds died en route to Maryland in 1665,[54] taking no part in transmitting these ideas to the Levant either.

Nayler's other leading adherent, Hannah Stranger, as well as her husband, John Stranger, repented of their actions and did not travel anywhere near the Ottoman lands. Neither did the 'resurrected' Dorcas Erbury, nor Jane Woodcock, nor any of Nayler's other followers at Bristol who were discharged before Nayler was taken to London.[55] Two of Nayler's followers, who did travel eastward, only made it as far as the Dutch Republic. Ann Cargill travelled to Amsterdam, where she apparently broke up any meetings where the Quaker leaders tried to speak against Nayler, but there is no mention of her preaching a messianic message.[56] Furthermore, Samuel Cater did not go to the Low Countries until 1669, which was years after the rise and fall of the Sabbatian movement.[57]

Quaker letters moved Nayler's story to Amsterdam, where the Quaker community was apparently 'severely shaken' by Nayler's messianic claims.[58] George Bishop wrote to Margaret Fell in the Dutch Republic:

[51]   Rosemary Moore, *The Light in their Consciences: Early Modern Quakers in Britain 1646–1666* (University Park: The Pennsylvania State University Press, 2000), 43.

[52]   Brailsford, *A Quaker from Cromwell's Army*, 157. Many of Nayler's contemporaries blamed her as well. Ralph Farmer, *Sathan Inthron'd in his Chair of Pestilence*, 22 concluded, 'Thus you see here, this Martha Simonds is a considerable person, For her Husband (who tis like knoweth her) tells her, *she was the chief leader* in this action'.

[53]   LSF Caton MS 391: Richard Hubberthorne to Margaret Fell, 1657.

[54]   Kenneth Carroll, 'Martha Simmonds, A Quaker Enigma', *Journal of the Friends Historical Society* Vol. 53 (1972), 51–2: another source says she was buried in Southwark. Regardless, she did not travel to the Ottoman Empire. For more on this elusive woman, see Carroll's article.

[55]   LSF *Joseph Joshua Green, Biography of Samuel Cater of Littleport in the Isle of Ely, Yeoman* (typewritten copy, 1914), 19, 20: Robert Crab was later imprisoned and died shortly after his release whereas Timothy Wedlock fell back into obscurity after 1656.

[56]   Popkin, 'Spinoza, the Quakers and the Millenarians', 123. According to Hull, the Nayler episode was 'accentuated in Amsterdam by Ann Gargill', who was apparently one of the women who had almost deified Nayler in Bristol. Hull, *The Rise of Quakerism in Amsterdam*, 273, 276,

[57]   Moreover, Cater had a falling out with Nayler shortly after the Bristol affair, writing to the former Quaker leader: 'Friend my soule is much burdened for thee … thou art departed', you are 'a stranger to the house of Christ'. LSF, Box A/4: Samuel Cater to James Nayler, 3 November 1656.

[58]   Popkin, 'Spinoza, the Quakers and the Millenarians', 122.

JN & his company (being released at Exeter) came into this Towne with full purpose and resolution to set up their Image & to breake the Truth in pieces & to bruise and tread towne, and beguile and devour the tender plants of the Lord in this his Vineyard as before was given forth; with which being overfilled and made drunk with the indignation of the lord they brought in JN on horseback, whoe rode with his hands before him; one reign of his bridle marth symds led and han string the other, and some went on his sides and Hannah's husband went bare before him & dorcas Erbury with a man of the Isle of Elyrod after, and thus they led him and thus he rode through the town, the women singing as they went holy, holy, holy, Hosannah and so past to the white hart a bad inne where they lay when they brought him first on the fifth month with hords following them (for the whole towne was mound) through the streets thither ... [59]

Bishop's words highlight the predominant view of Quakers who were not followers of Nayler. Even though many had spent time with him, they looked down on his actions in Bristol. William Edmondson, the founder of Irish Quakerism, had been convinced by Nayler himself in 1653, but he was filled with despair upon hearing of Nayler's messianic act:

But what added to my trouble, news was brought me of James Naylor's miscarriage. This came very near me, and brought me under great trouble of mind, so that I said in my heart, how shall I be able to stand through so many temptations and trials which attend me daily, since such an one as he is fallen under temptations? And I mourned in my spirit ... [60]

Because Nayler's actions were often used as a weapon to attack Quakerism throughout Europe,[61] most Friends would have been unlikely to spread word of what had transpired at Bristol.

Missionaries

While no one close to Nayler went to the Ottoman Empire, it has been asserted that 'a couple of Quaker missionaries' were preaching the imminent coming of the messiah in Jerusalem when either Sabbatai or Nathan was there.[62] The first attempt to carry the Quaker message to the shores of the Mediterranean began in 1657, so all of the missionaries would have been aware of Nayler's messianic entrance into Bristol. None of them, however, showed any form of messianic devotion to Nayler and very few of them knew him.

---

[59]   LSF Swarthmore MS I, 188: George Bishop to Margaret Fell, 27 August 1656.

[60]   LSF William Edmundson, *A Journal of the Life, Travels, Sufferings, and Labour of Love in the Work of the Ministry, of that Worthy Elder and Faithful Servant of Jesus Christ, William Edmundson* (Dublin, 1820), 68.

[61]   Brailsford, *A Quaker from Cromwell's Army*, 186.

[62]   Popkin, 'Christian Interest and Concerns about Sabbatai Zevi', 94; Goldish, *The Sabbatean Prophets*, 202.

The earliest Quaker mission to the Ottoman Empire included John Perrot, John Luffe, Mary Fisher, Mary Prince and Patrice Beckley, all of whom left England in 1657 and arrived in Livorno in July of that year. From Livorno, they went in different directions. Fisher, who can only be indirectly connected to Nayler,[63] travelled to Smyrna and Istanbul after apparently obtaining an audience with Sultan Mehmed IV in Adrianople in 1658. All accounts of her journey, however, fail to mention any interaction with Jews and, although Fisher may have been in Istanbul at the same time as Sevi, who stayed in the Ottoman capital for eight months around then, it is highly improbable that they ever met, had anything to do with each other or even knew of each other's existence.

John Luffe arrived in Smyrna in 1657 and spent time with Ottoman Muslims and Jews, who saw him as their wonder and 'gazing stocke'.[64] Yet he had little success in proselytising. Even if he had, Sevi was not in Smyrna during this period. Daniel Baker and Richard Scosthrop followed Luffe to Smyrna in 1661, when Sevi had returned to his hometown, and they spoke to the Jewish, Muslim and Greek Orthodox populations for three weeks. According to the missionaries, they were 'assured that their labour was not in vaine',[65] but even if they did interact with the Smyrnan Jews when Sevi was among them, Baker and Scosthrop had no ties to Nayler.

George Robinson was the man who apparently delivered the Quaker message in Jerusalem in 1657,[66] at the time that Nathan of Gaza was a student at a local yeshiva.[67] While it is very unlikely that the two men ever crossed paths, the Catholic friars in Jerusalem were made aware of Robinson's coming before he arrived, so he was not a regular pilgrim and would have attracted an unusual amount of attention, making it more likely that Nathan would have heard about him. That being said, Robinson had no connections to Nayler either.

John Stubbs and Henry Fell followed these Quakers eastward a few years later, setting out in 1661 for China and Prester John's country but ending up in Alexandria instead when they could not secure passage with any of the East India Company's ships.[68] Richard Bendige, the English consul at 'Grand-Cairo', was so annoyed with these missionaries that he advised his colleagues in Livorno not to let any more Quakers board ships to Alexandria. The ones who had made it, he wrote, had disbursed

---

[63] Fisher's husband, William Bayly, was one of the 22 Friends imprisoned in Exeter with Nayler and many of his adherents. Yet he did not marry Fisher until 1662, which was after her journey to the Ottoman Empire, and there is nothing to suggest that either Fisher or Bayly were followers of Nayler. See Brailsford, *A Quaker from Cromwell's Army*, 115; LSF Swarthmore MS I, 12.

[64] LSF Port MS 17.74: John Luffe to G.R., 10 October 1657.

[65] LSF Port MS 17.78. See Daniel Baker, *A Clear Voice of the Truth Sounded Forth* (London, 1662).

[66] Joseph Besse, ed., *A Collection of the Sufferings of the People called Quakers for the Testimony of a Good Conscience II* (London, 1753), 393.

[67] Goldish, *The Sabbatean Prophets*, 112.

[68] LSF Pamphlet Box L. 24: Bettina Lacock, 'Quaker Missions to Europe and the Near East 1655–1665' (undergraduate thesis, Birmingham University, 1950).

'[p]amphlets about the Streets in Hebrew, Arabick, and Latin, and if they had staid a little longer, it might have set them a burning'.[69]

When Stubbs and Fell were in Egypt, Sevi was possibly passing through. He had left Smyrna for Jerusalem in 1662, most likely travelling from Rhodes to Alexandria and arriving in Cairo, where he stayed for some time.[70] Unlike most of the missionaries, Stubbs was actually Nayler's friend. He visited Nayler in Exeter, wrote about Nayler's sufferings from Dublin and told Margaret Fell that 'James is pretty and deare to the whole household of god for ever'.[71] Even after the entrance into Bristol, Stubbs acted as Nayler's intermediary, interceding on his behalf with the Quaker leadership.[72] Stubbs, however, neither did nor wrote anything that demonstrated a messianic devotion to Nayler. Moreover, the overlap between him and Sevi was minimal even if he did leave Hebrew treatises for the Egyptian Jewry.[73]

After 1660, foreign missionary endeavours diminished due to concerns at home,[74] but a few Quakers continued to travel to the Levant. A 'John the Quaker' was known to have tried to convert the sultan in 1661,[75] and Charles Chillingworth, the English deputy consul in Livorno, wrote in 1666 that 'in the Sun from Constantinople came the worthy Mr Tho. Coke an English Gentleman, and John Filly of Dover an Emisary Quaker the latter intendeth for Florence and England'.[76] Such sparse references are a reminder of the possible importance of Quakers about whose journeys we know little to nothing.

Despite possible connections between the Quaker missionaries and the Sabbatian leadership, the English consul Thomas Bendish in Istanbul noted that the Quakers were generally 'censured and scoffed at, by Papist, Jew, and others of a strange faith'.[77]

---

[69]   Besse, *A Collection of the Sufferings of the People called Quakers II*, 420. Such a harsh reaction needs to be understood in its proper context: the printing of Arabic in the Empire was prohibited in order to protect the language of the Quran. See Matar, *Islam in Britain*,135.

[70]   Scholem, *Sabbatai Sevi*, 176–7.

[71]   LSF Swarthmore MS IV, 32: J[ohn] Stubbs to M[argaret] F[ell], 2 July 1656.

[72]   For instance, see LSF Swarthmore MS III, 152: J[ohn] Stubbs to M[argaret] F[ell], 10 June 1657.

[73]   In general, the Quakers wrote a considerable amount of material designed for the Jews. Missionaries often carried a Hebrew version of Margaret Fell's *A Loving Salutation to the Seed of Abraham* and, when John Perrot was examined before the Inquisition, he showed them a book 'he had written to the Turke and Jew' entitled *Immanuel, the Salvation of Israel*. Perrot also claimed to have 'given forth many books, both English, Latine, and French', including one to the governor of Livorno called 'the visitation of the Jewes written by G H' as well as 'a shorte paper to the Jewes' and 'a little booke which I was moved to write in the town to all the scattered Jewes throughout the world'. Considering the extensive Jewish connections between the Italian states and the Ottoman Empire, the Quaker message could have reached the Ottoman Jewry through this channel too, but this is a weak link at best. See LSF Port MS 17.76: J[ohn] P[errot] to E[dward] B[urroughs], 17 June 1657.

[74]   Barry Reay, *The Quakers and the English Revolution* (London: Temple Smith, 1985), 107.

[75]   Matar, *Islam in Britain*, 134: see J[ohn] P[errot], *The Blessed Openings of a Day of Good Things to the Turks* (1661).

[76]   TNA SP 98/6: Charles Chillingworth to Lord Arlington, 29 January 1666.

[77]   'Thomas Bendish's Report in the Calendar of State Papers', *Journal of the Friends Historical Society* Vol. 8 (1911), 168.

Gerardus Croese, the Dutch Reformed chaplain in Smyrna, similarly critiqued the Quaker missionaries for their lack of success:

> [T]he Men, except one or two, was so furnished by the Holy Spirit, as to be able to speak the Language of the Nation they came to wherever it was, without the help of an Interpreter, who himself seldom knew how to translate their sayings into the same sense and words, as they spake them, and besides might either, through mistake or on set purpose render them amiss. So all of them with great vehemence, Zeal and Industry, set about this work, but for all their care and pains could not do much good at it. Besides, that they also, which follow'd those that had gone afore them, altho they understood the Languages of the Countreys they went into well enough, yet made but small progresses. Wherefore in all those parts where these Preachers had travelled, at this day you shall find very few or no Quakers.[78]

Although Bendish and Croese were not on the friendliest terms with the Friends, the majority of Quaker missionary reports do not contradict their claims, suggesting that success was unlikely even among those who conversed freely with the Jews.

In sum, it must be conceded that there were, technically, enough possible associations to grant the possibility that one or more of the missionaries were acquainted with Nayler and travelled to areas where they could have shared his story with the Ottoman Jews and maybe even Sevi or Nathan. The amount of overlap, however, was minimal. None of these Quakers were at Bristol, were counted among Nayler's followers or spent a considerable amount of time with him or those closest to him. Although they were doubtlessly aware of what had transpired, the missionaries tended to view Nayler's actions negatively and would have avoided discussing the drama at Bristol. Even if they decided to tell the Jews about it, none of the missionaries spent a substantial period of time in close proximity to Sevi or Nathan. Furthermore, none of them mentioned meeting either the Jewish messiah or his prophet after the Sabbatian movement was known across Europe. Therefore, while there were Quakers in the Levant shortly before the outbreak of Sabbatianism, there was a very little possibility of the dissemination of the messianism associated with Nayler and his circle by the Quaker missionaries.

Merchants

The most extensive networks between England and the Ottoman Empire in the seventeenth century were mercantile. Were messianic ideas brought to the Ottoman Empire from England, and not by Quaker missionaries, they would most likely have been spread by merchants. There were merchants among Nayler's followers, and it has been suggested that they dispersed to the outposts of the Quaker commercial world in the Americas and the Levant after his punishment.[79] When Nayler was released from

---

78 Gerardus Croese, *The General History of the Quakers* (London, 1696), 167.
79 Popkin, 'Three English Tellings of the Sabbatai Zevi Story', 47.

prison, he lived with Rebecca Travers, who was known as a 'merchant's wife'.[80] While her husband, William Travers, was a merchant, he was only a tobacconist in London with no known connections to the Levant.

Robert Rich was another of Nayler's loyal followers who was a merchant. While Rich continued to hold strong messianic beliefs even after Nayler's imprisonment and fled to the outposts of the Quaker commercial world in 1658, he went westward to Barbados, from where he did business in New England. Rich did not go to the Ottoman realms.[81] Indeed, no known followers of Nayler were merchants who went to the Levant. Based on the movement of Nayler's followers, such as Rich and Simmonds, as well as the regular contact between Quakers in Bristol and their co-religionists in Barbados, Bermuda and Maryland, one would expect the messianic spirit to have been spread westward across the Atlantic instead of eastward across the Mediterranean.

Multiple historians have argued that ideas percolating in the European merchant colonies in the Levant influenced the young Sevi, and one scholar recently claimed that Sevi's beliefs 'did not grow simply out of Jewish soil but were irrigated by the apocalyptic enthusiasm of Protestant merchants who had carried their ideas to Smyrna from England, Holland and central Europe'.[82]

The port city of Smyrna was a hub of diverse Protestant merchants,[83] and Sabbatai possibly knew at least one English merchant through his father, Mordecai, who was a factor for the English – a profession common for the Jews at that time. Sabbatai's brothers followed in their father's footsteps and went into business with him, creating a channel in which news from England could possibly have reached the young Jewish messiah through his family.

In terms of the reports about Nayler, there was only a limited amount of time for transmission to have occurred along this path because Nayler entered Bristol in 1656, Sabbatai's father died in 1663 and, between the years, Sabbatai was away from Smyrna for extended periods of time. Word of Nayler's messianic actions could have come at a later date through Sabbatai's brothers, but Sabbatai engaged in frequent peregrination and, as time progressed, Nayler's actions became less and less newsworthy.

At least two historians have postulated that Mordecai was employed by English Quakers;[84] however, the specific sources proving this have not been provided and nothing in the historical record suggests that there were Quaker merchants in Smyrna. While Quaker missionaries often referred to merchants who helped them in their

---

    [80]   Fogelklou, *James Nayler*, 220.

    [81]   Nabil Matar, 'The Idea of the Restoration of the Jews in English Protestant Thought, 1661–1701', *The Harvard Theological Review* Vol. 78, No. 1–2 (1985), 67, 80; Fogelklou, *James Nayler*, 216, 225.

    [82]   Abulafia, *The Great Sea*, 482–3.

    [83]   Sonia Anderson, *An English Consul in Turkey: Paul Rycaut at Smyrna, 1667–1678* (Oxford: Clarendon Press, 1989), 17–18. For more on Smyrna, see Goffman, *Izmir and the Levantine World*.

    [84]   Goldish, *The Sabbatean Prophets*, 111; Popkin, 'Christian Interest and Concerns about Sabbatai Zevi', 94.

travels,[85] John Luffe wrote that none of the English in Smyrna stood with him.[86] A few years later, Baker similarly declared, 'there was no small stir among the men of their owne nation against them and their testimony and ... that equity and truth might have no entrance among them'.[87] These were the same merchants in Smyrna who had the English ambassador in Istanbul issue a warrant for the immediate removal of the Quaker missionaries. If these Englishmen were so adamant to expel the missionaries who were only there for a short time, surely a more permanent Quaker merchant would have attracted their attention as well.

Even statistically it is implausible that there were Quaker merchants in Smyrna. By 1660, there were at most 60,000 Quakers, which was less than one per cent of the total English population. Quaker recruits were found primarily among artisans and independent farmers whereas merchants and bankers were mostly against the Friends.[88] Considering the small English community in Smyrna had less than a hundred people, the likelihood of it including Quaker merchants was minimal.[89]

If, on the off chance, there were Quakers in Smyrna, the likelihood that they were followers of Nayler is even smaller based on demographics and the anti-Nayler views of many Quakers. Anyone with a serious devotion to Nayler would have been around him in the early 1650s, moving to the Levant most likely after his arrest in 1656. Extant diaries, correspondence and records relating to both the Quakers and the Englishmen in the Empire, however, provide no evidence of Nayler adherents in the city.[90]

---

[85]   For example, John Perrot wrote about a sympathetic French merchant in Livorno and George Robinson spoke of a French merchant who aided him in Accra. See LSF Port MS 17.76: J[ohn] P[errot] to E[dward] B[urroughs], 17 June 1657.

[86]   LSF Port MS 17.74: John Luffe to G.R., 10 October 1657.

[87]   LSF Port MS 17.78: Daniel Baker to unknown Quakers, 1661.

[88]   Damrosch, *The Sorrows of the Quaker Jesus*, 30.

[89]   The only indirect evidence for a Quaker population in Smyrna comes from the writings of the Dutch chaplains stationed there. Thomas Coenen, the Dutch pastor who witnessed the rise and fall of the Sabbatian movement firsthand, compared the Sabbatians and the Quakers in his treatise. Gerardus Croese, who replaced Coenen as the Reformed pastor in Smyrna, wrote a book titled *Historia Quakeriana* (Amsterdam, 1695) that described how Nayler entered Bristol with his followers in a manner that replicated Jesus' entry into Jerusalem and also mentioned the Quaker missionaries. The fact that two Dutch pastors in Smyrna published material that discussed the Quakers might suggest that the missionaries to the Levant garnered their attention or that there were Quakers in Smyrna more permanently because Croese himself wrote, 'I have had the Fortune of a long time to be familiarly acquainted and much conversant with these Men call'd Quakers, and that in many places; and have thus had occasion to know so much of them both from themselves and their chief Teachers'. Both Coenen and Croese, however, could have acquired their knowledge of and interest in the Quakers in the Dutch Republic, which was home to large communities of Friends. This appears to be the case because, according to Hull, *The Rise of Quakerism in Amsterdam*, 127, when Croese began to collect material for his text, he turned first to the Quaker Benjamin Furly in the Dutch Republic in 1690 followed by William Seawall who supplied him with a great deal of information. See Goldish, *The Sabbatean Prophets*, 62; Thomas Coenen, *Ydele Verwachtinge der Joden Getoont in den Persoon van Sabethai Zevi* (Amsterdam, 1669), 41; and Croese, *The General History of the Quakers*, 124, 167, 270.

[90]   The closest connection between Nayler and the English colony in Smyrna appears to come through Samuel Cater, the man who led Nayler's horse into Bristol. Years later, Cater travelled to the

It is possible that other members of the English colony in Smyrna who were interested in the events at Bristol, if only as observers, could have kept informed of the proceedings and told their Jewish associates about it. After all, the English merchants in Smyrna normally conducted their trade through Jewish middlemen, were in frequent communication with associates back home and made a point of entertaining English travellers, so they were relatively well informed about both Jewish affairs and news from England. But, once again, no surviving correspondence even hints at any communication about the Quaker messiah.

Problems with a Quaker Influence

The Quakers not only provided the largest number of extant pamphlets to, for and about the Jews, but Jewish–Quaker contact also occurred across the Atlantic and Mediterranean worlds. And the Jews were largely positive in response to the Quaker overtures to meet and discuss religious matters.[91] Notwithstanding the strong Jewish–Quaker connections, there is no proof that the Ottoman Jews knew about Nayler. Even if they did, knowledge of does not imply belief in. There are two primary problems with claiming that Nayler's act had a cross-religious influence on the outbreak of the Sabbatian movement.

The first relates to timing. Quakerism did not begin until the 1650s, Nayler did not ride into Bristol until 1656 and the messianic beliefs of Nayler's followers grew stronger closer to the Bristol affair. In short, Quaker messianism could not have affected Sevi before 1650, but Sevi claimed to have his first messianic vision as early as 1648 and was banished from Smyrna for his blasphemous behaviour sometime between 1651 and 1654.[92] If the young Sevi proclaimed himself the messiah in 1648, his messianic career began before both Nayler's and the advent of Quakerism. And why would we not accept the authenticity of Sevi's early claims? The year of Sevi's revelation was believed to be the Zoharitic messianic year 1648 based on gematria because the verse, 'In this year of the Jubilee shall ye return', equalled 5408, or 1648.[93] Not only are these early claims documented in the Hebrew historiography, but even

---

Dutch Republic with his close friend Giles Barnardiston, who was part of a prestigious family from Clare in Suffolk that may have included the merchant A. Barnadiston who worked with the Ottoman Jews in Smyrna. Cater, however, had a falling out with Nayler after Bristol and Giles Barnardiston did not become a Quaker until 1661, so the transmission would have had to occur after the falling out, and no documentation (including Giles Barnardiston's will) mentions an A. Barnadiston. See Giles Barnardiston, 'Abstract of Will of Giles Barnardiston', *Journal of the Historical Society of Friends* Vol. 7 (1910), 43–4.

[91]   For more on the interactions between Quakers and Jews, see Lesley Hall Higgins, 'Radical Puritans and Jews in England, 1648–1672' (doctoral dissertation, Yale University, 1967) and Stefano Villani, *I Primi Quaccheri e gli Ebrei* (Rome: Edizioni di Storia e Letteratura, 1997).

[92]   Scholem, *Sabbatai Sevi*, 151.

[93]   Abba Silver, *A History of Messianic Speculation in Israel: From the First through the Seventeenth Centuries* (New York: The Macmillan Company, 1927), 252.

the English merchants in Smyrna acknowledged them when they wrote in 1666 that Sevi had been 'pretending to bee their messiah' for 'neare 20 yeares'.[94]

Leaving aside the problem of timing, there is still the issue of content. How do the messianic claims, actions and ideas surrounding a lone Quaker in England from a decade earlier affect a Jewish messiah in the Ottoman Empire? There were no prophecies associated with Nayler that Sevi could have employed, and Sevi had his own messianic tradition from which to draw.

It is the similarities between Nayler and Sevi, not any sort of influence, which explain why their contemporaries grouped the two men together. Nayler's actions ran parallel to those of Sevi, and likening one detested religious group to another was a common rhetorical weapon.[95] Indeed, Christians used Sevi to attack the Quakers as radical millenarians who posed a serious political threat; the Jewish movement was a tool utilised by Christians against their fellow Christians.[96]

Linking Nayler and Sevi was also part of a larger trend in which the opponents of Quakers charged the Friends with secretly acquiring assistance from Jews, fraternising with Jews, using Hebrew in a suspicious manner and being Judaisers.[97] Even in North America, Cotton Mather commented on the 'dubious nature' of Quaker–Jewish affiliation.[98] All of these sources, from the German pamphlets and woodcuts to the writings of English Protestants on both sides of the Atlantic, have played a role in informing the views of historians who have remarked positively on the existence of 'Quaker Jews' and 'Jewish Quakers' as well as the possibility of connections between Nayler and Sevi.

Other Possible Christian Influences

Although Nayler did not have an impact on Sevi, there are numerous topics in Sabbatianism that suggest a cross-religious influence. For example, Nathan of Gaza's theology took on a larger Christian dimension as time progressed, especially around 1666, when he emphasised pure faith in a manner that is 'distinctly Christian in character'.[99] Meanwhile, Sevi's proclamation as *l'unico figliolo, e primogenitor d'dio* has conspicuously christiological overtones that apparently cannot be explained

---

[94] TNA SP 97/18/212: this unsigned letter dated 29 September 1666 to T. Dethick appears to be from S. Pentlow, J. Foley and T. Laxton.

[95] Shalev, 'Islam, Eastern Christianity, and Superstition according to Some Early Modern English Observers', 150.

[96] For more on the manner in which Christians connected Nayler to Sevi in order to criticise the behaviour of their fellow millenarian brethren, see Heyd, 'The "Jewish Quaker": Christian Perceptions of Sabbatai Zevi as an Enthusiast' as well as the subsection entitled 'Greate Hopes' in Chapter 4, 'A Jewish Messiah among Christians: The Evolution of European Perceptions of Sabbatai Sevi (1665–1666)'.

[97] Higgins, 'Radical Puritans and Jews in England', 188, 186.

[98] Higgins, 'Radical Puritans and Jews in England', 190: see Cotton Mather, *Magnalia Christi Americana II* (London, 1702), 257.

[99] Scholem, *Sabbatai Sevi*, 83.

phenomenologically and 'must be seen as an important case of direct influence of Christology on Sabbatai himself'.[100]

If not Nayler and the Quakers, then which Christians could have informed the Sabbatian movement? One scholar has noted that the widespread emergence of Christian millenarianism at the same time as Sabbatianism 'would be a striking coincidence indeed' if they were not connected,[101] and another scholar has suggested that the 'apocalyptic enthusiasm' of Fifth Monarchists and Rosicrucians was brought to Smyrna, where it influenced the young Sevi.[102] This chapter has sought to prove that Quaker messianism was not at the root of Sabbatianism, but what about these two other Protestant movements?

The Fifth Monarchy Men were an English millenarian movement that anticipated the imminent return of Jesus. They could not have had an impact on Sevi's early messianic claims because, like the Quakers, the Fifth Monarchists did not emerge until after 1648.[103] There were, however, more general apocalyptic fifth monarchy prophecies about 1666 that were circulated in England and across Europe that could have been used to support the growth of the general Jewish messianic excitement in that year. The Ottoman Jews may have known about the Christian expectations for 1666 because the Englishmen in Smyrna stated that 'the Jews themselves say that nothing made them so willing to believe as the Friar predictions on the yeere 1666'.[104] According to them, it was because 'the Xtians did foresee such strange revolutions which would happen in this yeare 1666 as likewise so that those Jewes in Xtian lande as with other pls [places]

---

[100]   Moshe Idel, *Messianic Mystics* (New Haven: Yale University Press, 1998), 205–6. According to Idel (202), Nathan even altered the popular Jewish attitude towards Jesus.

[101]   Katz, 'English Charity and Jewish Qualms', 262.

[102]   Abulafia, *The Great Sea*, 482–3.

[103]   It is important to differentiate between the ideas of fifth monarchism, which circulated across Europe, and the English Fifth Monarchy Men, which emerged following the king's execution in 1649 and never probably exceeded more than ten thousand followers. For more on the Fifth Monarchy Men, see Bernard Capp, '*A Door of Hope* Re-opened: The Fifth Monarchy, King Charles and King Jesus', *Journal of Religious History* Vol. 32, No. 1 (2008), 17–18 and Bernard Capp, *The Fifth Monarchy Men: A Study in Seventeenth-Century English Millenarianism* (London: Faber and Faber, 1972). In terms of more general fifth monarchy beliefs (some of which dovetailed with those of the English Fifth Monarchy Men), there were other apocalyptic pamphlets espousing similar ideas before 1648, such as Henry Archer's *The Personal Reign of Christ upon Earth* (1642) (This text was published multiple times, using both Henry and John as Archer's first name). While such texts that prophesied the imminent end of history were widespread – Archer's *The Personal Reign of Christ upon Earth* was printed six times in the decade after its release – and certain Englishmen in Smyrna may have been aware of such apocalyptic beliefs, there is no surviving evidence that suggests an influence on Sevi's early beliefs. The Christian theologians who wrote such pamphlets deduced from biblical calculations that great things were going to happen in 1650, 1656, 1666 and 1700, but there was no mention of the year 1648. Considering that some Jewish kabbalists anticipated the advent of the messiah in 1648 and that Sevi was a Jewish rabbi interested in the kabbalah who declared himself the messiah in that year, it seems much more plausible that his initial messianic claims were based upon his Jewish tradition.

[104]   TNA SP 97/18/211: A. Barnardiston, J. Adderley and N. Thurston to T. Dethick, 9 October 1666. Some Fifth Monarchy Men, such as John Rogers, too expected cataclysmic events to occur in the year 1666 because of the usage of the number 666 in Revelation.

were at one and the same time equally possessd with beliefe'.[105] If the Jews were aware of such Christian sentiments, the Sabbatians could have argued that the Christian prophecies about the year 1666 proved the legitimacy of their messianic movement. It therefore appears that, while Protestant apocalyptic beliefs did not inform Sevi's early messianic claims, Jewish and Christian eschatological enthusiasm did overlap and mutually reinforce each other in 1666 when Christian prophecies about the coming end dovetailed with similar Jewish expectations manifested in the Sabbatian ideology.

Unlike the Fifth Monarchy Men, the Rosicrucian movement was in full bloom before Sevi's first messianic claim in 1648. The Rosicrucian issue came to the fore in the Holy Roman Empire in the early decades of the seventeenth century after two manifestos, the *Fama Fraternitatis* (1614) and the *Confessio Fraternitatis* (1615), were printed in Kassel.[106]

Hundreds of tracts appeared in response to the publication of these manifestos, and some of them were millenarian in nature. Considering the growth of interest in Rosicrucianism among Protestants, could these ideas have reached the merchants in Smyrna, who spoke to their Jewish associates about them?

There is a possibility that this occurred, but it is unlikely. Rosicrucianism was most popular in central Europe whereas the Protestant merchants in the Levant came primarily from the Dutch Republic and England. While the Rosicrucian furore eventually gained attention in northern Europe, it originally seemed 'to arouse little or no public attention in Britain'.[107] The Rosicrucian manifestos were not even printed in English until 1652,[108] so it would be a stretch to suggest that Rosicrucianism spread from Germany to England where it was then moved to the Levant along mercantile channels to influence Sabbatai Sevi before 1648.

In regard to Dutch channels, the Rosicrucian manifestos were printed in Dutch in 1616 and Rosicrucian ideas were circulating in Batavia in 1629, so it is possible that Dutch merchants in Smyrna may have known about Rosicrucian matters. But even if they did, it is unlikely that the small Dutch community shared these ideas with the Jews in Smyrna in a manner that influenced the growth of their own eschatological

---

[105]   TNA SP 97/18/212: S. Pentlow, J. Foley and T. Laxton to T. Dethick, 29 September 1666.

[106]   The *Fama Fraternitatis* purported to be a booklet issued by a secret society that was founded by a medieval German monk named Christian Rosenkreuz who, after studying alchemy, mathematics and the kabbalah in the Levant and Africa, returned home and formed the Rosicrucian brotherhood. While both manifestos contained a 'hopeful and apocalyptic attitude', Rosenkreuz and the Rosicrucian brotherhood were completely fictional. It is important to note that 'Rosicrucianism', much like 'Quakerism', was a blanket term used by their adversaries, who had little regard for nuances. For more on the Rosicrucians, see the German scholarship by Carlos Gilly and Will-Erich Peuckert as well as Leigh Penman, 'Sophistical Fancies and Mear Chimaeras? Traiano Boccalini's Ragguagli di Parnaso and the Rosicrucian Enigma', *Bruniana and Campanelliana: Ricerche Filosofiche e Materiali Storico-Testuali* (Pisa and Rome: Fabrizio Serre Editore, 2009), 101–20.

[107]   Frances Yates, *The Rosicrucian Enlightenment* (London and New York: Routledge, 1972), 157. For the spread of Rosicrucianism in northern Europe, see Susanna Akerman, *The Rose Cross Over the Baltic: The Spread of Rosicrucianism in Northern Europe* (Leiden: E.J. Brill, 1998).

[108]   Yates, *The Rosicrucian Enlightenment*, 238.

expectations. Moreover, the debate over the Rosicrucian manifestos deteriorated quickly in Germany following the onset of the Thirty Years' War, so the interest in Rosicrucianism was dying while the Jewish messianic beliefs were growing.

Although it has been said that the Rosicrucian manifestos contained a 'powerful prophetic and apocalyptic note',[109] they were tame compared with the millenarian pamphlets that circulated in England and other places at that time. The Rosicrucian movement was devoid of messianism, concentrating instead on the chiliastic impulse of a circulation of wisdom in the last days. More important, the manifestos contained no prophecies, statements of doom or biblical calculations claiming that Jesus' return was going to occur between 1648 and 1666.[110]

It seems more likely that both Rosicrucianism and Fifth Monarchism informed the messianic actions of James Nayler rather than Sabbatai Sevi. In fact, unlike with the Jewish messiah, one can actually trace a path between both movements and the Quaker messiah through Martha Simmonds. Simmonds, *'the chief leader'* in Nayler's messianic entrance into Bristol,[111] first encountered Nayler through his writings that were sent to London to be published by Simmonds' brother: the infamous printer Giles Calvert.[112] Calvert also sold the Rosicrucian manifestos and printed many Fifth Monarchist works.[113] If Simmonds read Nayler's treatises that were published by her brother, why would she not have read the Rosicrucian manifestos and the Fifth Monarchist pamphlets as well? A Fifth Monarchist influence is quite plausible considering that they preached the second coming of Jesus in the very year that Nayler replicated Jesus' actions at the urgings of Simmonds. Thus, it is more likely that the Rosicrucian manifestos and, especially, the Fifth Monarchist prophecies had an

---

[109]   Yates, *The Rosicrucian Enlightenment*, 66.

[110]   While it is highly unlikely that Rosicrucianism was at the root of the Sabbatian movement, it appears that Sabbatianism influenced the eighteenth-century neo-Rosicrucian movement. In the 1780s, an organisation that would become known as the Asiatic Brethren of St. John the Evangelist in Europe emerged. Although they did not use the Rosicrucian name, they were part of the neo-Rosicrucian current. An important member of this organisation was Baron Thomas von Schonfeld, who had been a follower of the Sabbatian movement and incorporated certain Sabbatian doctrines into the order's teachings. For more, see Christopher McIntosh, *The Rose Cross and the Age of Reason: Eighteenth-Century Rosicrucianism in Central Europe and its Relationship to the Enlightenment* (Leiden: E.J. Brill, 1992) and Jacob Katz, *Jews and Freemasons in Europe 1723–1939* (Cambridge, Mass.: Harvard University Press, 1970).

[111]   Farmer, *Sathan Inthron'd in his Chair of Pestilence*, 22.

[112]   Neelon, *James Nayler*, 124–5. For more on Giles Calvert, see Altha Terry, 'Giles Calvert, Mid-Seventeenth Century English Bookseller and Publisher: An Account of his Publishing Career, with a Checklist of his Imprints' (master thesis, Columbia University, 1937); Altha Terry, 'Giles Calvert's Publishing Career', *Journal of Friends Historical Society* Vol. 35 (1938), 45–9; Mario Carrichio, *Religione, Politica e Commercio di Libri nella Rivolzione Inglese. Gli Autori di Giles Calvert, 1645–1653* (Genoa: 2003); and Mario Carrichio, 'News from the New Jerusalem', in Ariel Hessayon and David Finnegan, eds, *Varieties of Seventeenth and Early Eighteenth Century English Radicalism in Context* (Aldershot: Ashgate, 2009), 69–86.

[113]   Thomas Willard, 'The Rosicrucian Manifestos in Britain', *Bibliographical Society of America Papers* Vol. 77 (1983), 493.

impact on the Quaker messiah in England than either did on the Jewish messiah in the Levant.

If none of these Protestant groups influenced the Sabbatian movement, how does one explain the aforementioned Christian elements of the Sabbatian ideology? The Sabbatian leaders probably acquired their Christian ideas from a quasi-Christian source located much closer to them. Both the messiah and his prophet lived in cities with former Conversos who had lingering Christian values and a long history of eschatological speculation.[114]

Many members of former Converso families played a prominent role in the Sabbatian movement, especially in Smyrna, where several of Sevi's close friends and fellow students from former Converso families had a significant impact on him and his messianic interests. Sevi spoke Spanish rather than Turkish, sang Spanish *romanceros* and knew a great deal about Christianity.[115] In regard to Nathan, his teacher in Jerusalem was the former Converso Jacob Hayyim Tzemah, who wrote a book that contained visible echoes of the Christian kabbalah.[116] Indeed, 'it is now known that Christian messianism, Portuguese Marranism, and Sabbateanism were inextricably tied together'.[117] One should therefore turn first to the former Conversos in the Ottoman Empire to find the Christian background that informed Sevi and Nathan's beliefs and not to fringe Protestant movements a continent away.

The opening woodcut of James Nayler and Sabbatai Sevi points to a cross-religious connection that looks good on paper, especially seventeenth-century paper, but it had and still has very little substance. By examining the networks surrounding Nayler, this chapter has reached two conclusions. The first half of the chapter has added to the historiography by employing recently discovered Italian and Dutch sources to prove that the Bristol affair was known about to a degree not previously recognised. The story surfaced in handwritten correspondence and publications from one side of Europe to the other.

Very few of the 20 biographies of Nayler discuss how he was understood and presented in other countries. Only recently has Stefano Villani, the authoritative historian of Quakers in Italy, written about the Italian diplomatic dispatches on

---

[114] Messianism was a prominent part of the beliefs of Judaising Conversos. For more, see Yerushalmi, *From Spanish Court to Italian Ghetto: Isaac Cardoso*, 304–5; A.Z. Aescoly, 'David Reubeni in the Light of History', *The Jewish Quarterly Review* Vol. 28, No. 1 (1937), 37; Stephen Sharot, 'Jewish Millenarianism: A Comparison of Medieval Communities', *Comparative Studies in Society and History* Vol. 22, No. 3 (1980), 399; and Gerson Cohen, 'Messianic Postures of Ashkenazim and Sephardim', in Marc Saperstein, ed., *Essential Papers on Messianic Movements and Personalities in Jewish History* (New York and London: New York University Press), 205.

[115] Barnai, 'The Sabbatean Movement in Smyrna', 114–16; Goldish, *The Sabbatean Prophets*, 1–2. For more on Sabbatai singing Spanish *romanceros*, see Gad Nassi, 'Meliselda: The Sabbatean Metamorphosis of a Medieval Romance', *Los Muestros* Vol. 48 (2002), 38–41.

[116] Idel, *Messianic Mystics*, 206.

[117] Barnai, 'Christian Messianism and the Portuguese Marranos', 119. Barnai (121) does, however, admit that 'these possibilities, even probabilities, are insufficient to remove all doubts whether there was a Christian or Marrano messianic influence on Sabbatai Zevi himself or members of his circle'.

Nayler. But even Villani was not aware of the existence of printed Italian avvisi reports about the Quaker leader. By comparing and connecting the English, Latin, Italian and Dutch narratives about Nayler across religious, national and professional boundaries, this chapter has sought to provide insight into cross-religious representations, transnational transmission and the entangled relationship and disjunctures between correspondence and publications in the dissemination of news.

Shifting away from the concrete, the second half of the chapter has argued that the messianism associated with Nayler did not make it to the Ottoman Empire and certainly not in a manner that affected the outbreak of the Sabbatian movement. The idea that Protestant millenarianism in Europe influenced the emergence of Sabbatianism was popularised two decades ago and, although it is not taken as seriously now, this problematic argument has still been repeated in recent texts by historians.

Although this chapter has reached a negative conclusion, this outcome has two broader implications. First, it points to an asymmetrical relationship between the Ottoman Empire and the European states. Stories from the Levant were disseminated broadly in gazettes and pamphlets across Europe whereas accounts of events in Europe may have been transmitted to the Ottoman realms in letters or by word of mouth by merchants, diplomats or other travellers; however, the lack of a news industry in the Ottoman domains meant that the European reports never reached as large a population in the Empire.

Secondly, it highlights a similar asymmetrical relationship between the Jewish and Christian worlds. While the seventeenth-century Jewry may have been small and dispersed compared with the expanding Christian states, the transmission of the Nayler narrative shows that incidents in Christendom did not have as direct or as strong an influence on Jewish religious thought compared with the impact that incidents in the Jewish world had on Christian perceptions of their own religious tradition. Indeed, as the next two chapters demonstrate, news and rumours related to the emergence of the Jewish Sabbatian movement would be spread quickly and broadly throughout Catholic and Protestant lands, affecting the hopes and expectations of Christians across Europe and beyond.

Chapter 3
Who Sacked Mecca? The Life of a Rumour (1665–1666)

In the autumn of 1665, English readers received extraordinary news through several channels simultaneously. Although specific details varied, all the reports contained the same basic story: the city of Mecca, Islam's holiest site, had been sacked. Moreover, some of the earliest versions added a second element of no less astonishing significance: the enormous army attacking Mecca, they claimed, was none other than the Lost Tribes of Israel.

As many seventeenth-century Jews and Christians were aware, the Bible records that 10 of the tribes of Israel were taken into captivity by the Assyrian King Salmanassar, after which they are not mentioned again. Since the messianic promises of the Old Testament were often addressed to the descendants of Abraham, Isaac and Jacob, prophetic exegetes had long speculated that these 10 'Lost' Tribes must have been preserved in some distant corner of the globe from which they would suddenly re-emerge to play a crucial role in ushering in the long-sought messianic period. As Jewish messianic hopes intermingled with Christian millenarian ones, some seventeenth-century Christian millenarians entertained heightened expectations regarding the return of the Lost Tribes as well.[1] And since Islam had played a significant role in Christian apocalyptic beliefs since its advent in the seventh century,[2] the sack of Mecca pit two of the greatest eschatological agents against one another. Nor was it lost on contemporary observers that this momentous event was taking place in the last months of 1665, on the eve of 1666 – the year that certain Christians speculated would be the final one in history.

Such sensational news could only travel the long distance from Mecca to London by passing through numerous and very different networks. In each, this was a story of the greatest possible significance, though its interpretation altered radically as it passed from one cultural milieu to another. In the Muslim world, the destruction of the holiest place in Islam represented an enormous disaster, but one predicted in certain apocalyptic scenarios. Within the Jewish world, the sudden appearance of a great host

---

[1]   See the numerous works by Richard Popkin, such as his 'Some Aspects of Jewish–Christian Theological Interchanges in Holland and England 1640–1700', in Johannes van den Berg and E.G. van der Wall, eds, *Jewish–Christian Relations in the Seventeenth Century: Studies and Documents* (Dordrecht: Kluwer Academic Publishers, 1988), 3–32.

[2]   For more, see Paul Alexander, *The Byzantine Apocalyptic Tradition* (Berkeley: University of California Press, 1985).

of their long-lost brethren seemed to inaugurate the final phase of an age-old story of national redemption. Christians, although standing on the sidelines, could perceive in this conflict variously the victory of one ancient eschatological adversary over another or, from a philo-Semitic millenarian perspective, the triumph of a recently acquired eschatological ally over the eastern antichrist. Despite all of these differences, this event had one thing in common for Jews, Christians and Muslims: it was of such outstanding significance that, if proved true, it must be part of God's divine plan for the imminent end of the world.

Given the power of these associations, it is scarcely surprising that the rumour of an attack on Mecca surfaced in correspondence, gazettes and pamphlets written in Hebrew, Italian, German, French, Dutch, Latin and English. Nor, perhaps, is it surprising that, in passing through so many cultural filters en route to England, this narrative was transformed almost beyond recognition. Indeed, one might claim that the least important part of the legend is a proper historical account of what actually transpired in the Arabian Peninsula in or around 1665 – where the Lost Tribes did not, in fact, suddenly reappear at Mecca, and where the holy city of Islam was not utterly plundered and destroyed. Rather, this is the history of how the nightmare of one people was fused with the dreams of another, before arriving in a variety of Christian cultures predisposed to accept and elaborate it further by deep-seated hopes and fears of their own. The story about how this news was created cannot be disentangled from the story of how it was transmitted. In what follows, we shall therefore need to attend even more closely to the cultural prisms through which the tale passed to reveal its tenuous historical origins.

Conception and Birth of a Threat to Islam

Mecca's role in the history of Islam begins with the biblical patriarch Abraham, who is said to have built the *Kaaba* with his first-born son Ishmael to hold a black stone given to him by an angel.

Muhammad too claimed to be visited by an angel, who told him to recite the divine word of God. When Muhammad shared his revelation with the people of Mecca, the city's leaders turned against him. Forced to flee to Medina, Muhammad gained a mass following before returning to Mecca, where he destroyed the pagan idols and declared the city to be the holiest site in Islam. From that time onwards, only Muslims have been allowed to enter Mecca, in order to protect it from outside religious influences. To this very day, Muslims are reminded of the importance of Mecca when they perform the daily *salat*, the prayer facing the Kaaba. If they are able, all Muslims are also expected to undertake the *hajj*, a pilgrimage to Mecca, at least once in their lives.[3]

---

[3]   See P. Bearman, T. Bianquis, C.E. Bosworth, E. van Donzel and W.P. Heinrichs, eds, *Encyclopaedia of Islam* (Leiden: Koninklijke Brill, 2006).

The centrality of Mecca has made it a point of vulnerability, and periodic threats have promoted fears of Mecca's destruction to resurface intermittently in the Islamic world. Even the Quran speaks of one such incident in the *sura*, or chapter, that reads,

> Hast thou not seen how thy Lord dealt with the army of the Elephant?
> Did he not cause their stratagem to miscarry?

The 'army of the Elephant' referred to a Christian king from Ethiopia who attacked Mecca in 570, the year that Muhammad was born.

In the twelfth century, this sura was reinterpreted as a prophetic reference to another Christian who threatened Mecca, Reynald of Chatillon. A French Christian who served in the second crusade, Reynald became the prince of Antioch during the years that Saladin was the protector of Mecca and Medina. In 1182, Reynald launched an attack on an oasis along the pilgrimage route to Mecca, which led to hostilities between the Islamic and Christian kingdoms. During this conflict, Reynald sent a raiding party against the two holy cities with plans to seize 'the Prophet's body'.[4] The Muslim forces captured Reynald's raiding party a day's march away from Medina, but Saladin was criticised for failing to protect the hajj.[5]

Mecca was not only threatened by foreign Christians; it was also ravaged by Arab tribes. Two centuries before Reynald, in 930, the Qarmatians attacked Mecca, massacring the pilgrims and carrying away the black stone in the Kaaba.[6] Less than 50 years later, in 976, Mecca was besieged again, this time by the Fatimid *caliphate*. Then, in 1309, the Sultan Oljeytu converted to Shi'ism and planned to invade Mecca, exhume the remains of the first two *caliphs* (Abu Bakr and Umar) and relocate them to a new shrine that he was building.[7]

While the region stabilised in the early modern period with the Ottoman conquest, Bedouin marauders posed a continued nuisance to both the city and pilgrims on hajj.[8] In 1632, a local dynasty re-established itself in Yemen and defeated the Ottoman troops, who fled to Mecca where they ran amok, mutinied and occupied the city. Such havoc triggered more Bedouin raids.[9]

The onslaught of the Ottoman army was the largest attack prior to 1665, when the rumour of Mecca's destruction appeared, which suggests that the story may not have

---

[4] Bernard Hamilton, 'The Elephant of Christ: Reynald of Chatillon', *Studies in Church History* Vol. 15 (1978), 97–104.

[5] Hamilton, 'The Elephant of Christ', 104n.

[6] Said Amir Arjomand, 'Islamic Apocalypticism in the Classic Period', *The Encyclopaedia of Apocalypticism II* (New York: Continuum, 2000), 272.

[7] Judith Pfeiffer, 'Confessional Polarization in the 17th Century Ottoman Empire and Yusuf Ibn Ebi Abdu'd-Deyyan's *Kesfu'l-Esrar fi Ilzami'l-Yehud ve'l-Ahbar*', in Camilla Adang and Sbaine Schmidtke, eds, *Contacts and Controversies between Muslims, Jews, and Christians in the Ottoman Empire and Pre-Modern Iran* (Wurzburg: Ergon Verlag Wurzburg, 2010), 42, 54n.

[8] Suraiya Faroqhi, *Pilgrims and Sultans: The Hajj under the Ottomans 1517–1683* (London: I.B. Tauris, 1996), 22.

[9] Faroqhi, *Pilgrims and Sultans*, 67.

been inspired by an actual event. One scholar has asked, 'Why do historians need a historical event as a peg to hang a legend on?'[10] Maybe this question should be posed in relation to the rumoured sack of Mecca, especially since there does not appear to be any historical basis to the tale.

If it was not based on a factual attack, Islamic eschatological expectations growing in this period of instability may have been at the root of the rumour instead. Mecca's central role in Islamic thought coupled with a legacy of threats against the city meant that certain Sunni and Shi'i apocalyptic sequences came to contain a siege on Mecca. It is not a coincidence that Mecca was threatened by an Ethiopian Christian army and that a whole family of eschatological traditions exist about a Christian attack on Mecca and Medina that will include a two-pronged assault from the Ethiopians in the south and the Byzantines in the north. According to these predictions, both holy cities of Islam will be sacked and, if not totally destroyed, at least seriously damaged.[11]

In the mid-1660s, there was a *mahdist* movement in Mecca.[12] The outbreak of a large following around a mahdi, an Islamic saviour expected to appear near the end of time, may have sparked apocalyptic tension that led to the resurgence of prophecies about the anticipated destruction of Mecca. Such conjecture could have been transformed into, what was perceived as, fact during transmission. It may have been one, all or none of these factors that manifested themselves in the earliest written version of the rumour, printed in a Venetian avviso on 25 April 1665:

> It is written from Livorno that a ship arrived from Smyrna in a short number of days, which confirmed that there were uprisings in Babylon and that an army of numerous Arabs had gone to Mecca, robbed the place, and carried away the corpse of Muhammad together with the treasury of the city. This shocked all those of the country, and they desire to get the confirmation for the news.[13]

Based on reports from the Levant,[14] the worst case scenario for Muslims had occurred. Mecca had been plundered, and Muhammad's body removed. The apocalyptic prophecies were being fulfilled.

---

[10]   C.F. Beckingham, 'The Achievements of Prester John', in Charles Beckingham and Bernard Hamilton, eds, *Prester John, the Mongols and the Ten Lost Tribes* (Aldershot: Variorum, 1996), 2.

[11]   David Cook, *Studies in Muslim Apocalyptic* (Princeton: The Darwin Press Inc., 2002), 261. For more, see Cook's subsection on 'Attitudes towards Cities', 254–68.

[12]   Goldish, *The Sabbatean Prophets*, 37, notes that a mahdist movement occurred in Mecca at this time, but he does not elaborate upon it.

[13]   BNCF Codd Magliabechiani XXV, 743, 81: Venice, 25 April 1665. The original read, 'Vien scritto da Livorno esser cosa capitata Nave da Smirne in pochi giorni, quale confermava, che in Babbillonia vi sussero delle sollevazioni, e che un'Esercito numeroso d'Arabi fosse andato alla Mecca, svaligiato quell luogo, portato via il Cadavere di Maeometto assieme col Tesoro, che cosa si trovava, e messo in costernazione tutto quell Paese, del che con desiderio se n'attende la confermazione'.

[14]   As discussed later in this chapter, in the section titled 'Circulation among Christians', the evidence suggests that the rumour actually emerged in the Ottoman Empire, despite the fact that most of the remaining sources were written in Europe.

This story, however, only involved Arabs. It was an 'Esercito numeroso d'Arabi' (numerous army of Arabs), who had committed these acts.[15] While this was a startling piece of news to those who heard it, a still more sensational version was yet to emerge.

Rebirth due to Jewish Prophecies

After the initial account of the Arab army sacking Mecca appeared in the Venetian press, the rumour went into a period of gestation; it was not mentioned again for almost half a year. It might even have died out completely were it not for the prophecies of Nathan of Gaza, whose letters outlining the theological foundations of Sabbatianism were circulated across the Jewish world.[16]

While certain Jewish messianic scenarios included the return of the Lost Tribes of Israel alongside the messiah, the Lost Tribes did not play a major part in Nathan's imagination at the beginning of the messianic movement; he only indirectly spoke of them when he disclosed the course of future events. In one of his letters that was disseminated widely, Nathan claimed that Sabbatai would proceed to the river Sambatyon beyond which the Lost Tribes were believed to dwell. Then, in 1672, he would return 'mounted on a celestial lion; his bridle will be a seven-headed serpent and fire out of his mouth devoured'.[17] At this sight, all the nations and kings would bow before him and the 'ingathering of the dispersed shall take place'.[18] Although the Lost Tribes were only implied, popular imagination seized upon this detail with an enthusiasm unforeseen by the prophet. Instead of the future date of 1672 envisioned by Nathan, his followers stated that at least two of the Lost Tribes, the 'Sons of Reuben and Gad', were to appear that year on the tenth or twentieth of *Shebat* (16 January 1666).[19]

When Raphael Joseph, the head of the Egyptian Jewry in Cairo, heard about Nathan's prophecies, he sent a series of emissaries to investigate. Sabbatai had stayed with Raphael Joseph in Cairo prior to 1665, and the head of the Egyptian Jewry was curious to learn more.[20] One of Raphael Joseph's emissaries was his brother, Hayyim Joseph, who stayed in Gaza and regularly reported the latest goings-on to Raphael.

---

[15] BNCF Codd Magliabechiani XXV, 743, 81: Venice, 25 April 1665.

[16] Idel, *Messianic Mystics*, 198.

[17] Nathan of Gaza to Raphael Joseph, September 1665, as translated and quoted in Scholem, *Sabbatai Sevi*, 273.

[18] Nathan of Gaza to Raphael Joseph, September 1665, as translated and quoted in Scholem, *Sabbatai Sevi*, 274.

[19] Scholem, *Sabbatai Sevi*, 352. Emanuel Frances, in a note about one of his poems in *Sevi Muddah*, stated that Nathan prophesied that the tribes of Gad and Reuben would conquer Palestine that year. See Scholem, *Sabbatai Sevi*, 353n.

[20] Martin Jacobs, 'An Ex-Sabbatean's Remorse? Sambari's Polemics against Islam', *Jewish Quarterly Review* Vol. 97, No. 3 (2007), 351.

Hayyim's accounts were 'wrapped in clouds of legends'.[21] He stated that Nathan tried to board a ship to join Sabbatai, but after Nathan was stopped by a storm numerous times, an angel came to him in a pillar of fire and told him to remain in Gaza because the Jews' redemption was at hand. Then, when Nathan was in the wilderness, the prophet Jehu (son of Hanani) appeared and confirmed the angel's message. He said that part of his tribe would soon arrive in Gaza.[22]

Raphael was excited by this news and passed along copies of Hayyim's letter with his own additions to their other brother, Solomon, in Livorno. One letter told the story of the rabbi Benjamin of Jerusalem, who had accompanied Nathan into the wilderness where 'Jehu b. Hanani' appeared to them and announced that two emissaries from the Lost Tribes would come from beyond the river Sambatyon within two months.[23] Nathan's distant prophecies were therefore given tangible form with immediate dates by the Jews who moved them along their familial and mercantile networks from Gaza to Egypt and across the Mediterranean. From Italy, the correspondence was spread across Europe, prompting excitement about the imminent arrival of the Israelites.

The prophecies about the Lost Tribes became so important to Sabbatian believers that at least three editions of Nathan's prayerbook printed in Europe had full-copperplate engravings on the frontispiece that referenced the ancient Hebrews. The cover contained two pictures. The top half showed Sabbatai as a king seated on his throne; the bottom half showed 12 bearded men and an oversized thirteenth. The entire episode was portrayed, in other words, as the 12 tribes of Israel joining forces with the messiah.[24]

Neither Nathan nor those close to him were concerned with Mecca or even aware of its rumoured destruction. But the prophecies of the Lost Tribes emanating from their circle were conflated with the rumour of the sack of Mecca when these stories reached Alexandria. This port city was central to pilgrimage routes to Mecca, overland trade routes to Gaza and shipping routes to Livorno and Venice. It is therefore not surprising that the first traceable instance of this conflation occurred in Alexandria, from which it then entered Europe via Livorno.

A Jew from the Egyptian port city told the Jewish scholar Raphael Supino in the Italian port city in June that the *bassa* of Alexandria and a 'king of Arabia' had made a journey to Mecca with 60,000 men. When the Arabs got closer, they 'learned that the city was besieged and partly conquered by unknown people who called themselves Israelites'.[25] The Arabs engaged the Israelites in battle but were soundly defeated. Upon hearing this news, the Jews in Alexandria sent emissaries to Mecca to find out the

---

[21]   Scholem, *Sabbatai Sevi*, 353.

[22]   Scholem, *Sabbatai Sevi*, 353. This prophecy was repeated twice in letters from Raphael Joseph to his brother Solomon Joseph in Livorno, which Frances apparently drew upon. It is a reference to Jehu, son of Hanani, mentioned in the biblical books of Kings and Chronicles.

[23]   Scholem, *Sabbatai Sevi*, 353.

[24]   Scholem, *Sabbatai Sevi*, 526.

[25]   Scholem, *Sabbatai Sevi*, 345.

truth, and the emissaries returned with confirmation that the army indeed comprised the long-lost Hebrews.[26]

The report was wrong. There was not a massive army of Israelites in the Arabian Peninsula. There may, however, have been a large, armed population in the desert because the yearly pilgrimage from Egypt to Mecca included at least 50,000 participants who were protected by a detachment of up to 300 janissaries, 100 cavalrymen and as many as 6 cannons.[27] This raises an interesting question. Could Nathan's prophecies have affected Sabbatian believers to such an extent that they were ready to interpret any news, possibly even a report about the pilgrimage to Mecca that was twisted and changed during transmission, as proof that their salvation was at hand?

In Livorno, Raphael Supino accepted the truthfulness of the story of the Israelites at Mecca and wrote about it in letters to his friends, including Jacob Sasportas, the rabbi of the Sephardic community in London.[28]

As the rumour spread throughout the European Jewries, it took on more layers. In Casale Monferrato, it was reported that the Jewish army had conquered Mecca and now planned to march against the Germans and Poles who had persecuted their brethren.[29] This particular version of the rumour was influenced by the following prophecy made by Nathan in a letter that was disseminated across the Jewish world:

> A year and a few months from today, he [Sabbatai] will take the dominion from the Turkish king without war ... and all the kings shall be tributary unto him, but only the Turkish king will be his servant. There will be no slaughter among the uncircumcised, except in the Ashkenazi lands.[30]

Popular imagination, it seems, seized upon Nathan's prophecy and mixed it with the rumour to create the next objective for the mythological army.[31] In Egypt, the Arab army was reconfigured into a Jewish army. In Italy, the mythological army was given another mission: the Israelites who had defeated the Muslims would now wage war against the Christians.

---

[26] Scholem, *Sabbatai Sevi*, 344–7.

[27] Faroqhi, *Pilgrims and Sultans*, 46, 54, 69.

[28] Scholem, *Sabbatai Sevi*, 344–7, 336.

[29] There was a long history of beliefs in the Lost Tribes returning to punish Christians for years of Jewish oppression, and such expectations prepared the minds of certain Jews to accept this story. For more on earlier Jewish beliefs, see Micha Perry, 'The Imaginary War Between Prester John and Eldad the Danite and its Real Implications', *Viator: Medieval and Renaissance Studies* Vol. 41, No. 1 (2010), 21, 18.

[30] Letter from Nathan of Gaza to Raphael Joseph as translated and quoted in Goldish, *The Sabbatean Prophets*, 76–7.

[31] Liebes, *Studies in Jewish Myth and Jewish Messianism*, 96. Scholem, *Sabbatai Sevi*, 347–51, relied on Hebrew and German sources, which encouraged him to see the version in Casale Monferrato as coming to Italy from the Balkans through Vienna. Considering the origins of the Venetian avvisi reports and the strong Jewish connections between the Italian peninsula and the Levant, it seems just as likely that it came from the Ottoman Empire across the eastern Mediterranean through Venice. This is more plausible because the accounts from Vienna stated that the leader of the military force was Jeroboam whereas, like the rumour in Casale Monferrato, the Venetian news items did not supply a name.

In the summer of 1665, the tale proliferated and grew increasingly fantastical. Letters spread among the Jews soon spoke of Mecca's total destruction. Because the Lost Tribes were expected to return with the advent of the messiah, these stories confirmed and bolstered the eschatological hopes, especially when they were followed by reports of Sabbatai Sevi's messiahship.

By the end of 1665, the letters dispatched from the Egyptian Jewry were more careful; they claimed that no more news was coming from Mecca.[32] But it was too late. The news had already spread across Europe and caused great excitement among certain communities. The Jews of Vienna, for instance, had made 'a publick Jubile' for the 'success of their Brethren in *Asia* against the Turk'.[33]

Circulation among Christians

Although the rumour only referred to populations of Jews and Arabs in the Ottoman Empire, it was disseminated most broadly by Christians in Europe due to their advances in printing, the heritage shared with the Jews, and the history of conflict between the Islamic lands and Christendom. With the incorporation of the sacred Hebrew texts into the Christian canon, the belief in the return of the Lost Tribes had passed into the Christian eschatological sequence ages earlier, and many seventeenth-century Christians expected the Israelites to re-emerge shortly before the end of the world.

As early as the seventh century, the conflict between Christians and Muslims led a significant number of Christians to believe in the advent of a saviour figure who would avenge their losses. The Syriac apocalypse of Pseudo-Methodius, for example, told the story of the rise of Islam and interpreted the Muslim military victories as a sign of the coming end. When there was no hope left, a Christian king would come forth, deliver the Christians, punish the Muslims and devastate Egypt, Hebron and Arabia.[34] The Toledo Letter, circulated five centuries later in 1186, claimed that one day a great conquest of the Islamic world would take place in which Mecca, Basara, Baghdad and Babylonia would be 'utterly destroyed'.[35]

As the success of the crusades diminished, the circulation of prophecies about the 'imminent end of Muslim rule' found a larger audience in Christendom.[36] In 1219, one of the leaders of the fifth crusade, the pope's legate cardinal Pelagio Galvani, who was dispatched to Damietta in Egypt, had a Latin translation made of the prophecy of Hannan, son of Agip. According to this prophecy, an army from the west would

---

[32]   Scholem, *Sabbatai Sevi*, 347.

[33]   *London Gazette*, 5 March 1666.

[34]   Alexander, *The Byzantine Apocalyptic Tradition*, 20–21.

[35]   Bernard McGinn, *Visions of the End: Apocalyptic Traditions in the Middle Ages* (New York: Columbia University Press, 1979), 152.

[36]   McGinn, *Visions of the End*, 154.

conquer Egypt while the king of the Abbisi would invade Arabia, attack Mecca and scatter the bones of Muhammad.[37]

The two-pronged attack from Europe on one hand and Abbisi or Abyssinia (modern-day Ethiopia) on the other has obvious parallels to the aforementioned Islamic eschatological scenario. Could these Christian and Islamic apocalyptic expectations have intertwined to form the basis of the rumour of the sack of Mecca that included the removal of Muhammad's body? At the very least, these beliefs foreshadowed it, preparing Christians and Muslims to anticipate the occurrence of such an event.

The Ottoman domination of the Middle East led to a new round of prophecies that incorporated the changing geo-political environment. Some Christians believed that the sultan was the antichrist and the Ottoman Empire was the biggest threat to the spread of Christianity across the entire globe: it was an obstacle that needed to be removed before the second coming of Jesus.[38] While Martin Luther tied the end of history to the rise of the 'Turk',[39] the English chiliast Thomas Brightman specifically stated that the 'obliteration of the Turks' would begin in 1650 and be accomplished by 1695 or 1696.[40]

For certain Christian millenarians, the destruction of Ottoman power was a vital step in the eschatological sequence, and they developed military scenarios in which they played out their apocalyptic fantasies. In *The Worlds Great Restauration* (1621), Henry Finch wrote that the 12 tribes of Israel would defeat the Ottomans in battle before converting to Christianity and returning to the holy land.[41] In *Iewes in America* (1650), John Dury proclaimed:

> God [will] call the ten Tribes to march toward the place of their inheritance: the Caraits their brethren will be leaders of them on their way, and so their march may be, as *Manasseh Ben Israel* saith, to make their Rendezvous in *Assyria;* and on the other side, the Jewes that are Pharisees, may make their Rendezvous from *Arabia* and other neighbouring places, and out of all *Europe* into *Egypt* [in] ... *the company of two Armies,* which both shall look towards *Jerusalem.* Then will the great battaile of Harmageddon be fought.[42]

---

[37]   Bernard Hamilton, 'Continental Drift: Prester John's Progress though the Indies', in Charles Beckingham and Bernard Hamilton, eds, *Prester John, the Mongols and the Ten Lost Tribes* (Aldershot: Variorum, 1996), 243–4. For more on cardinal Pelagio Galvani, see Joseph Donovan, *Pelagius and the Fifth Crusade* (Philadelphia: University of Pennsylvania Press, 1950).

[38]   Hill, 'Till the Conversion of the Jews', 14.

[39]   See Gordon Rupp, 'Luther Against "The Turk, the Pope, and The Devil"', in Peter Brooks, ed., *Seven-Headed Luther: Essays in Commemoration of a Quincentenary 1483–1983* (Oxford: Clarendon Press, 1983), 255–73.

[40]   Paul Christianson, *Reformers and Babylon: English Apocalyptic Visions from the Reformation to the Eve of the Civil War* (Toronto: University of Toronto Press, 1978), 105.

[41]   See Henry Finch, *The Worlds Great Restauration* (London, 1621).

[42]   Thorowgood, *Iewes in America*, An Epistolicall Discourse of Mr. Iohn Dury.

Christian eschatological hopes such as these set the stage for the propagation of tales about the Lost Tribes.

Unlike these earlier stories, those in the seventeenth-century gazettes were not based on biblical exegesis. They were news. They were proof that the ancient prophecies were coming true. As early as 1645, the *London Post* reported that the Jews were collecting themselves into one body to return to Palestine. Two years later, a news pamphlet entitled *Doomes-Day; or the Great Day of the Lords Judgement proved by Scripture* (1647) announced that the Jews were assembling in Asia under the leadership of Josias Catzius for the final overthrow of the antichrist.[43] There were similar reports in 1648.[44]

Although these accounts would have heightened messianic and millenarian excitement, the fact that they were printed so regularly may have made people sceptical. In 1645, for example, Ralph Josselin wondered doubtfully if the story in the *London Post* could be true.[45] Five years later, Thomas Thorowgood conceded in his *Iewes in America* (1650) that this is 'an age much enclining to Enthousiasmes and Revelations.'[46]

The regularity with which such news items had been proven false should have soured Christian views about the tale of the Lost Tribes sacking Mecca in 1665; but this rumour was so widespread and so close to the predictions of some theologians that certain people, including scholars, were evidently inclined to believe it. Sure as they were that it was only a matter of time before the biblical prophecies would be fulfilled, they found these expectations validated by the news that the ancient Hebrews had returned and attacked the Muslim holy city. The end times were finally upon them, and they shared the story in haste.[47]

Italian Catholics at Home and Abroad

Due to the numerous networks that connected the Ottoman Empire to the Italian peninsula, the tale of Mecca's destruction first entered Christendom through Catholic

---

[43]   Katz, *Philo-Semitism and the Readmission of the Jews to England*, 102.

[44]   Hill, 'Till the Conversion of the Jews', 27.

[45]   According to Hill, Ralph Josselin was preoccupied with the conversion of the Jews and thought that it might begin in 1654. Hill, 'Till the Conversion of the Jews', 20.

[46]   Thorowgood, *Iewes in America*, The Epistle Dedicatory.

[47]   Because the rumour aligned so closely with these Christian hopes and the extant evidence is predominantly written by Christians, one may want to argue that this was a purely Christian phenomenon, created and spread by them. It is true that there is no proof that Ottoman Muslims knew of the rumour (most likely because there are not many sources from this period that would contain such information), and there is only circumstantial evidence that the Ottoman Jews did (in the form of Supino's report that he heard the story from a Jew from Alexandria). However, avvisi and gazettes tended to be reliable in their claims of where their information originated, and Scholem and van der Wall both stated that the evidence suggests the European Christians may have exaggerated the stories about the Israelites in their letters, but they did not make them up. Thus, past scholarship coupled with analysis make it relatively safe to conclude that the rumour emerged somewhere in the Ottoman Empire while the new evidence proves that it did so independently of the Sabbatian movement.

Italy. The Venetian avviso editor acquired his information from the Levant; Raphael Supino in Livorno heard about the attack from an informant from Alexandria; and the Venetian senators were at least partially told about it from one of their representatives in Istanbul, the Italian diplomat Giovan Battista Ballarin.[48]

Like most diplomats, Ballarin negotiated trade agreements, interceded on behalf of his country's merchants in day-to-day affairs and wrote regular dispatches for his government. On 18 March 1666, Ballarin closed his weekly diplomatic dispatch by stating that, although the report he was about to relate might seem fictitious or exaggerated, it was essential to understanding the manner in which the power of the inhabitants of Barbary was growing due to something new that was emerging there. For some months, a Jew 'di assai bella apparenza, ma di proffonda dottina' (of beautiful appearance, but [also] of great learning) in Arabia had given himself the 'titolo di messia' (title of messiah) and gained a considerable number of Jewish followers.[49]

While the rest of Ballarin's letter was dedicated to Sabbatai Sevi, who had recently been imprisoned near Istanbul, Ballarin's comment about Barbary and Arabia is noteworthy because it suggests ties to the army of the Lost Tribes, which was supposed to be active in this region. The other sources that incorrectly located Sabbatai in these places only did so when they claimed that he was leading a Jewish military force that comprised the Israelites. One therefore wonders if Ballarin was hearing and passing along a version of Sabbatai's history that mixed the Jewish messiah's biography with elements of the fictitious attack on Mecca.

Ballarin was the only known European in the Levant even indirectly to reference the rumour. Although many European merchants and diplomats in the Ottoman Empire thought that Sabbatai and his followers were important enough to discuss in their regular correspondence, none of them told the tale of Mecca's destruction, even though it was published widely in sources throughout Europe. Diplomatic and mercantile letters were considered fairly reliable, and the fact that neither contained references to the sack of Mecca suggests one of two situations: either the merchants and diplomats were more poorly informed about a supposed event in their state of residency than their countrymen back home or they were much better informed and knew that it was completely fantastical and did not warrant any attention. But if the latter was the case, why did not a single European stationed in the Levant simply say so in one of their dispatches sent to places where the rumour ran rampant?

The narrative of the plundering of Mecca first appeared in a Venetian avviso long before both Ballarin's report and Nathan's prophecies. Yet the rise of the Sabbatian movement inspired a wave of messianic excitement that ultimately led the same avviso to reprint the story four months later with a significant addition. On 8 August 1665, the avviso stated that the Arabs at Mecca were now joined by a 'numero grandissimo' of Jews from the surrounding region who had appointed a leader that they called their

[48] Scholem gives his name as Giambattista Ballarino. For more on Scholem's analysis of this letter, see Scholem, *Sabbatai Sevi*, 447–55.

[49] ASV Senato Dispacci Ambasciatori Constantinopoli, F. 150, 19b–21: Giovan Battista Ballarin to the Venetian doge and senate, 18 March 1666.

king.[50] A few weeks after that, another news item in the avviso confirmed the account. The Arabs at Mecca were with a 'numero infinito' of Jews.[51] They had robbed and burnt Mecca before returning to their country with their rich booty.[52]

The addition of the Jews to the rumour reflects the influence of the Sabbatian movement: the story of an Arab army attacking Mecca resurfaced with a contingent of Jews during the period in which the Venetian Jewry were hearing similar tales about the Israelites from their brethren. This should not be surprising. Venice had both a blossoming news agency and a vibrant Sephardic community with connections throughout the Levant. Not only do the descriptions of the Jews as a *numero grandissimo* and *numero infinito* resonate with the Lost Tribes' boundless image, but a great number of Jews following a man that they called king also contains aspects of the Sabbatian believers' views of Sabbatai.

The Jews of Venice would have put the reports from the Venetian avvisi together with the letters from the Egyptian Jewry in a manner that furthered the belief that the ancient Hebrews had indeed returned. In other words, the press in a Catholic state was indirectly confirming the unbelievable news from the Ottoman Jews that their messianic hopes were coming true.

People, especially Catholics, may have been sceptical about the items in the Venetian avvisi because they were based on hearsay. When news did not have a specific origin, avvisi editors described their sources as oral communication.[53] The accounts of the Jewish army were full of such references, including 'viene ancora replicato' (it is still repeated), 'parla costantamente dell'invasione fatta' (the invasion is constantly spoken about) and 'pervenuta da piu parti la confermazione' (to be confirmed from many quarters), which suggests that they came from overheard conversations in Venice rather than from the Ottoman Empire or elsewhere.[54]

The following month, the rumour regressed back to its original form *sans* Jews. On 5 September, the Venetian avviso editor acquired letters from the Levant which confirmed that, alongside the rebellion in Babylon, 'un corpo d'Armata d'Arabi' (troops of an Arab army) sacked Mecca and took Muhammad's body away along with the city's treasure.[55] Like the original report and unlike the other two that followed it, this one only mentioned an Arab force, listed its sources and referred to a rebellion in Babylon. These similarities substantiate the argument that the middle versions, containing an enormous Jewish contingent, came from oral sources influenced by the increased attention to Sabbatian prophecies because the two printed avvisi that did not mention the Jews actually provided the origins of their reports.

---

[50]   BNCF Codd Magliabechiani XXV, 743, 84: Venice, 8 August 1665.

[51]   BNCF Codd Magliabechiani XXV, 743, 93: Venice, 29 August 1665.

[52]   BNCF Codd Magliabechiani XXV, 743, 93: Venice, 29 August 1665.

[53]   Villani, 'Conscience and Convention', 81–2.

[54]   BNCF Codd Magliabechiani XXV, 743, 84: Venice, 8 August 1665 and BNCF Codd Magliabechiani XXV, 743, 93: Venice, 29 August 1665.

[55]   BNCF Codd Magliabechiani XXV, 743, 94b: Venice, 5 September 1665.

The treatment of the rumour in the Venetian avvisi sheds light on the relationship between rumour, knowledge and credibility during this period. There was no standard progression in the stories in the avvisi: the Arab force gained a Jewish element before returning to its original form. There was also no discussion of this incongruity. The editor was not concerned with discrepancies; he just printed the news items without comment, listing the sources or lack thereof to allow the readers to judge each one for themselves.

The changing nature in the Venetian accounts dovetailed with a similar process in the Jewish transmission. Among the Jews, the rumoured amount of tribesmen grew with every version. Tens of thousands of men became hundreds of thousands, and several sources gave the exact number of 1,100,000. Moreover, very few of the early narratives referenced the standard Jewish belief that the Israelites would converge upon the holy land.[56] There were direct parallels in the Venetian avvisi: the number of Jews increased from great to infinite, there was no logical sequence in the army's composition and they merely went home after conquering Mecca – there was no mention of Jerusalem or the holy land.

It is likely that these similarities were more than just coincidence. If the rumour of the Lost Tribes conquering Mecca resulted from a cross-religious conflation, then the original report in the Venetian avviso explains why the early stories spread among the Jews did not refer to the Israelites regaining possession of the holy land, which was supposed to be one of their primary roles in the eschatological sequence. Based on news of a military force attacking Mecca and returning home, the earliest change to the tale was a simple substitution of the populations; the storyline remained the same.

This shows a complex interaction between the developments of the two sets of narratives. On the one hand, rumours about the Lost Tribes circulating among the Jews were combined with a story of an Arab sack of Mecca to create a mixed Jewish–Arab force in the later Venetian avvisi. On the other hand, the lack of eschatology in the original version that only spoke of an Arab attack caused the tales about the Lost Tribes at Mecca disseminated among the Jews to leave out their end time's mission of restoring the holy land, which might have been expected had the story simply been invented.

The rumour turned up in the Turin press as well, although at a significantly later date. On 25 February 1666, an avvisi editor in Turin printed information contained in letters from Istanbul 'ricevutesi da buona mano' (received from a good hand), which stated that the sultan had sent the bassa of Aleppo to the Arabian Peninsula.[57] The Jewish and Arab peasants united in revolt against the bassa, creating a force of 140,000 combatants who slaughtered the 'Turks'.[58]

While the Venetian avvisi provided abstract figures, such as a *numero grandissimo* and a *numero infinito*, the Turin avviso gave the army an actual number: there were

---

56 Scholem, *Sabbatai Sevi*, 343.
57 SA Segreteria di Stato, Avvisi, 148: Turin, 25 February 1666.
58 SA Segreteria di Stato, Avvisi, 148: Turin, 25 February 1666.

140,000, in this case, militiamen. The Italian press was not generally concerned with the possible millenarian ramifications of this news; however, the figure of 140,000 could be drawn from two sources of apocalyptic significance. First, it could be related to the 144,000 Jews mentioned in the book of Revelation, a clear connection to Christian eschatological expectations. Secondly, and far more unlikely, it could be the faint echoes of the rumour's Islamic origin. The fourteenth-century Muslim scholar Ibn Kathir summarised the Islamic apocalyptic scenario as follows:

> The Antichrist will be allowed to appear at the end of times after the conquest of Constantinople by the Muslims. He will appear first of all in the Jewish quarter of Isfahan, followed by 70,000 Jews, all armed, as well as 70,000 Tartars and people from Khurasan.[59]

The first part of the prophecy had come true. Constantinople had fallen to the Muslims. Now the Turin avviso was telling of the emergence of 140,000 attackers, half of whom were Jewish and half of whom were not – just like those spoken of by Ibn Kathir. These parallels could of course merely be coincidence, or these details found in the Italian avviso could be the last visible traces of the Islamic root of the rumour. Even if they were coincidental, such a narrative would have at least partially conformed to Muslim hopes regarding the coming end.

Unike its Venetian counterpart, the Turin avviso added that 'la sollevatione degli Hebrei, & Arabi' (the uprising of the Jews and Arabs) had always been held to be fantastical until the letters from Istanbul confirmed it.[60] The editor in Turin had obviously been aware of the story for quite some time but had been sceptical. He was waiting for a reliable account before printing it.

As the rumour travelled around Italy, it also continued to change in Venice. A year after it first appeared, on 10 April 1666, a Venetian avviso stated that letters from Ragusa (modern-day Dubrovnik) had no more information '[d]el progresso dell'Armi Ebraiche nella Palestina' (of the progress of the Jewish army in Palestine).[61] Then, it shifted to letters from Istanbul: no one was certain where Sabbatai was on his journey from Alexandria, but his imminent arrival was worrying the Jews of the Ottoman capital.[62]

Even though the avviso had no new information about either story, the editor still deemed both newsworthy. Furthermore, he placed the two reports, containing drastically different truth values, together. The former was about a fictitious military force; the latter was about a real person. The new army in Palestine was most likely believed to be the same one from Mecca because the avviso noted that there was no

---

[59]   As translated in Jean-Pierre Filiu, *Apocalypse in Islam*, trans. M.B. DeBevoise (Berkeley: University of California Press, 2011), 39. In *Les Signes due Jour Dernier*, Ibn Kathir stated that Medina and Mecca would escape harm – unlike other Islamic prophecies.

[60]   SA Segreteria di Stato, Avvisi, 148: Turin, 25 February 1666.

[61]   SA Segreteria di Stato, Avvisi, 148: Venice, 10 April 1666.

[62]   SA Segreteria di Stato, Avvisi, 148: Venice, 10 April 1666.

more known about the army, an obvious continuation of a previous report.[63] The growth of the Sabbatian movement was therefore continuing to affect the rumour's content in the Venetian press: the army was now completely devoid of Arabs, it was moving towards the holy land and it was discussed in relation to the Jewish messiah. The news items in the press in Catholic Italy were becoming more eschatological in nature.

The stories in the printed Italian avvisi were supplemented by at least one pamphlet. In 1666, a short Italian publication known as 'A Christian Report from Constantinople regarding Shabbetai Sevi' presented the history of the Jews' messiah and his followers. During its discussion of the messianic movement in the Levant, the pamphlet spoke of 'the mysterious army on the march from the East' to assist the messiah in his mission.[64] Although not central to the storyline, the unnamed mysterious army still featured as part of the Sabbatian plot.

Jews valued the tales of the return of their long-lost brethren because it supported their messianic hopes. It did not matter where the Lost Tribes were; it mattered that they had finally appeared to save the Jews. This explains why the letters among the Jews told of sightings of the ancient Hebrews in numerous places besides Mecca.

Italian Catholics, on the other hand, never mentioned the Lost Tribes in any of their writings. Their understanding of the rumour was in direct opposition to the Jews'. They valued it more for its claims about the destruction of the Muslims' holy city. For them, it did not matter who had sacked Mecca; it mattered that Mecca had been sacked. Such an event would have had political and religious significance considering the history of Venetian–Ottoman conflict.

Avvisi were important sources of news and, in them, the rumour was placed among other items of a political and military nature. In Venice, it was put next to a story about a rebellion in Babylon. In Turin, it was said that the Ottoman authorities were ignoring the threat against Islam's holiest site because they were 'più inclinata a gl'apparecchi contro Christiani' (more focused on their preparations [for war] against the Christians).[65]

Since this was a period of decline for the Ottoman Empire in which it was losing land to its enemies and Istanbul was almost starved by the Venetians, the propagation of this tale among the Italians points to the role that Mecca played in their imagination. Many Christians may not have had accurate knowledge about the city. Some did not know that Muhammad was not buried there; his grave was located in Medina. The Venetian navy, however, had shown that the Ottomans' political and economic capital

---

[63] A letter from Jews in Alexandria to their brethren in London claimed that the Lost Tribes had come through Prester John's country in Africa, marched for a year and a half to take Mecca and were currently believed to be on the borders of Canaan. This shows that another source in circulation provided similar routing for the army. See JTS *The Congregating of the Dispersed Jews* (London, 1666).

[64] S. Simonsohn, 'A Christian Report from Constantinople regarding Shabbethai Sevi (1666)', *Journal of Jewish Studies* Vol. 12, No. 1–2 (1961), 36. For more on this pamphlet as well a copy of it in its original Italian, see this article.

[65] SA Segreteria di Stato, Avvisi, 148: Turin, 25 February 1666.

was vulnerable, and the decimation of Mecca would have demonstrated that Islam's religious heart on the far side of the Empire was not invulnerable either.[66] On the contrary: the destruction of Muhammad's tomb was believed by some to signal the imminent demise of the Ottoman Empire.[67]

In the Multi-religious Dutch Republic

In the Low Countries, the rumour multiplied. The multitude of networks, especially mercantile ones, extending out of the Dutch Republic in all directions meant that information flowed to Amsterdam via oral and written communication from the Levant and the Americas. Much of the news found its way to the economic centre of Amsterdam, where the main financial institutions were located within a few minutes' walk of each other. It is no coincidence that, when coffeehouses became popular, they were built in this area.[68] Moreover, a tolerant policy and an abundance of economic opportunities led Amsterdam to develop into a pluri-confessional city that became the most important meeting point for seventeenth-century Jewish and Christian intellectuals.[69] Christian scholars interested in Hebrew and biblical scholarship turned to the Jewish immigrants who had Christian backgrounds, a broad European education and spoke several languages. Public debates began as early as 1606 and rabbis, such as Menasseh ben Israel and Elijah Montalto, wrote polemics that were disseminated throughout the Christian world.[70]

In the Dutch Republic, the rumour first appeared in cheap, throw-away pamphlets that told of the miraculous return of the Lost Tribes alongside other wondrous tidings. According to one of them, the walls of the second temple were slowly rising from the ground in Jerusalem too. As early as July 1665, barely a month after Nathan had anointed Sabbatai in Gaza, the *Herstelling van de Joden* (1665) elaborated upon the rumour with all manner of corroborative detail. The Ottoman janissaries, it reported, had been slaughtered by miraculous intervention: after firing their weapons, they had

---

[66]   As far as Venetian–Ottoman warfare was concerned, the years between 1661 and 1666 were quite peaceful, so at most it was the legacy of conflict and the proximity of the looming Ottoman military threat that made this story newsworthy in Catholic Italy. For a political history of Venice and the Ottomans, see the chapter on 'The Turco-Venetian War (1646–1653) and the Turmoil in Istanbul' in Kenneth Setton, *Venice, Austria, and the Turks in the Seventeenth Century* (Philadelphia: American Philosophical Society, 1991).

[67]   Ingrid Maier and Daniel Waugh, "'The Blowing of the Messiah's Trumpet": Reports about Sabbatai Sevi and Jewish Unrest in 1665–67', in Brendan Dooley, ed., *The Dissemination of News and the Emergence of Contemporaneity in Early Modern Europe* (Farnham: Ashgate, 2010), 144.

[68]   De Vries and van der Woude, *The First Modern Economy*, 148.

[69]   J.L. Price, *Dutch Society 1588–1713* (Harlow: Pearson Education, 2000), 204.

[70]   Theodor Dunkelgrun, "'Neerlands Israel': Political Theology, Christian Hebraism, Biblical Antiquarianism, and Historical Myth* (Leiden: Koninklijke Brill, 2009), 224–6. For more on the Dutch Jewry, see Jozeph Michman, ed., *Dutch Jewish History: Proceedings of the Fifth Symposium of the History of the Jews in the Netherlands* (Jerusalem: Institute for Research on Dutch Jewry, 1991) and Chaya Brasz and Yosef Kaplan, eds, *Dutch Jews as Perceived by Themselves and by Others: Proceedings of the Eighth International Symposium on the History of the Jews in the Netherlands* (Leiden: Brill, 1998).

watched in horror as their projectiles turned around in mid-flight and came down upon them instead. The Lost Tribes killed two-thirds of the Ottoman force, took control of the road to Mecca and allowed none to pass. It was even said that the sultan dreamt that an Israelite had taken the crown from his head – a passing reference to Nathan's prophecy about the sultan becoming Sabbatai's servant perhaps.

The German *Wahrhafftes Conterfey* (1665), printed a short time later, told a similar story. It claimed that 300,000 tribesmen killed all the non-Jewish inhabitants of Mecca, put the tomb of Muhammad on a cart and left the city in ruins.[71] The number 300,000 requires further comment because it is the same figure that is used in a narrative about a sixteenth-century Jew named David Reubeni. A century before Sabbatai's birth, Reubeni arrived in Europe, saying that he was sent by his brother Joseph, the king of the Lost Tribe of Reuben in Arabia. Joseph, who was in command of 300,000 Israelite warriors,[72] wanted his emissary to forge an alliance with the pope to 'overcome all Muslims in war and subjugate Mecca'.[73] Reubeni's claims were taken seriously: he was granted an audience with the pope and his story was recounted across Christendom. One wonders if this version of the rumour swirling around Mecca incorporated the reference to the 300,000 tribesmen from Reubeni's sixteenth-century account, reframing it in terms of other material that was circulating in 1665.

While sources such as the *Wahrhafftes Conterfey* and the *Herstelling van de Joden* confirmed the news that the Jews of Amsterdam were receiving in letters from their brethren, the Dutch Sephardim may have been sceptical of the reports because similar pamphlets had contained accounts of Jewish women giving birth to pigs and elephants. Sasportas and sceptics like him who first heard the rumours of the Jewish movement from printed broadsides demanded confirmation from more reliable sources, such as newspapers.[74]

Dutch Christians were also sceptical. The Protestant scholar Petrus Serrarius had been interested in the Lost Tribes for 20 years, ever since his friend Menasseh ben Israel had become excited over Antonio de Montezinos' discovery of the Israelites in the jungles of South America. Yet Serrarius did not initially trust the news of the Lost Tribes sacking Mecca. He wrote in September that he refused to believe that Mecca had been besieged because he had read nothing about it in the Dutch gazettes.[75] Like Sasportas, he did not think that the pamphlets were sufficiently trustworthy. They were known to tell sensational stories. Gazettes, on the other hand, were seen to carry

---

71 Van Wijk, 'The Rise and Fall of Shabbatai Zevi as Reflected in Contemporary Press Reports', 15, 17.

72 Parfitt, *The Lost Tribes of Israel*, 208. For more on David Reubeni, see Aescoly, 'David Reubeni in the Light of History'.

73 Benite, *The Ten Lost Tribes*, 114.

74 Maier and Waugh, "'The Blowing of the Messiah's Trumpet"', 150.

75 Van Wijk, 'The Rise and Fall of Shabbatai Zevi as Reflected in Contemporary Press Reports', 15, 17.

more sober information supplied by respectable compatriots on the scene, such as ambassadors, clergymen, soldiers and merchants.[76]

Serrarius did not have to wait long for the confirmation that he desired. In January 1666, two prominent Dutch gazettes, the *Oprechte Haerlemse Saterdaegse Courant* and the *Oprechte Haerlemse Dingsdaegse Courant*, presented the following news item:

> The Jews and Arabs have sacked the grave of Mohammed in Mecca and conquered many places; so that the Turkish Court had offered to vacate Alezandria Tunis and other Places; but they desired the whole Holy Land.[77]

The gazettes followed the Italian avvisi, not the Dutch pamphlets, in claiming that it was an Arab–Jewish force in the Arabian Peninsula. But eschatological expectations, rarely found in the Italian press, infiltrated the gazettes in the form of the comment that the army 'desired the whole Holy Land'. Thus, even the most secular Dutch version of the rumour, containing strictly Arab and Jewish combatants, already had an implicit apocalyptic component.

The items in the Dutch gazettes were printed at around the same time as the one in the Turin avviso, which was at a significantly later date than the reception of letters about the sack of Mecca by the Amsterdam Jewry as well as the publishing of the tale in the Dutch pamphlets and the Venetian avvisi. The varying paths and speeds of the rumour demonstrate a complex interaction between distance and scepticism: even if the story arrived sooner along some channels, certain people refused to print it and others refused to believe it until they heard it from sources that they trusted.

The *Hollandtze Mercurius* was a Dutch newsbook that was considered a reliable news provider similar to the Dutch gazettes. It too published numerous items about the sack of Mecca, provided many details and explicitly connected the destruction of Mecca to the Sabbatian movement. Even in the register, the newsbook grouped the two together, placing the account of the 'Jootze gepretendeerde Messias Nathan Levi' (the Jews' supposed messiah Nathan Levi) next to that of the 'Joodtze op.-tocht in Arabien' (the Jews marching across Arabia).[78]

In October 1665, the *Hollandtze Mercurius* conflated the two stories by making Sabbatai the leader of the attack: a Smyrnan native (that is, from the same city as Sabbatai himself) was followed by thousands of Jews who went along the Red Sea to Mecca where they intended to destroy Muhammad's tomb. From there, they were going to Jerusalem, where he was to be crowned king before they triumphantly

---

[76]   Van Wijk, 'The Rise and Fall of Shabbatai Zevi as Reflected in Contemporary Press Reports', 15, 17.

[77]   *Oprechte Haerlemse Saterdaegse Courant* and the *Oprechte Haerlemse Dingsdaegse Courant*, 23 and 26 January 1666, as translated and quoted in van Wijk, 'The Rise and Fall of Shabbatai Zevi as Reflected in Contemporary Press Reports', 21, 21n.

[78]   ULL Pieter Castelyn, ed., *Hollandtze Mercurius, Vervatende het Gepasseerde in Europa: Voornamentlijck in den Engelze ende Nederlantschen Oorlog, in 't Jaer 1666* (Haarlem, 1667), Register I.

proceeded to Istanbul.[79] The rumour was therefore continuing to evolve during transmission: the key figure of the Sabbatian movement, not the hopes or prophecies relating to it, was now part of the storyline.

Six months later, the siege of Mecca was once again front and centre in the *Hollandtze Mercurius*. Similar to the developments in the Venetian avviso, the time elapsing between the printing of the two stories facilitated the development of the rumour in the Dutch newsbook. In these new accounts, the mythical army was no longer led by Sabbatai, but by a Jew named Giorobaon. The newsbook proceeded to tell of his rise to power. In the kingdom of Elal in the state of Aden in Arabia, a man named Giorobaon had gained mass support and, with the assistance of his followers, taken Medina and Mecca.[80]

The name Giorobaon would resurface in connection with the rumoured attack on Mecca in numerous Dutch and English sources. Who was this man? Although the prefix 'Gio' sounds Italian, the Italian press only stated that the Jews and Arabs were following a man 'intitolatolo con nome di Re' (they called their king).[81] The avvisi never mentioned his name. Jewish letters spread throughout Europe from Vienna, however, claimed that a Jew from Aden named Jeroboam had risen up against the Ottomans, defeated them with a mighty army of Jews and Arabs and taken over at least 70 cities.[82] The name 'Jeroboam' therefore seems to have been added to the tale by Jews somewhere east of Vienna but not south from Italy.

Shared with Christians, the name was changed into 'Giorobaon' in the Low Countries, where some versions of it lost an 'o' and became 'Giorbaon'. This path of transmission seems to have had an oral component because, although 'Jeroboam' and 'Giorobaon' are spelt very differently, they sound similar. This suggests that 'Giorobaon' was a mishearing of 'Jeroboam'. While both names were used in Dutch pamphlets and newsbooks, only 'Giorbaon' is found in the English sources; it was solely the final version of the name that made it across the English Channel.

Like the army that he supposedly led, the man named Jeroboam, Giorobaon or Giorbaon was completely fictitious; yet the choice of this name does not appear to be random. It was probably based on the biblical figure Jeroboam, whose story is found in the books of Kings and Chronicles. According to these texts, Jeroboam was a mighty man of valour whom God promised to make king of Israel. After Solomon's death, the 10 northern tribes revolted and invited Jeroboam to become their ruler. Jeroboam accepted, and his new kingdom waged numerous wars against Rehoboam, his successors and the kingdom of Judah. Jeroboam even brought an army of 800,000 men against them. Due to Jeroboam's sins, God told him that his house would be destroyed and Israel would be rooted out of the land and scattered beyond the river – the events that gave rise to the creation of the Lost Tribes of Israel.

---

[79]   ULL Pieter Castelyn, ed., *Hollandtze Mercurius, Behelzende de Gedenckweerdichste Voorvallen in 't Jaer 1665* (Haarlem, 1666), 148: October 1665.

[80]   ULL Castelyn, *Hollandtze Mercurius*, 52–3: March 1666.

[81]   BNCF Codd Magliabechiani XXV, 743, 84: Venice, 8 August 1665.

[82]   Scholem, *Sabbatai Sevi*, 347.

In seventeenth-century Europe, the biblical account of Jeroboam was known to be linked to Jewish eschatological expectations due to the writings of Menasseh ben Israel. The *Hope of Israel* stated:

> Observe, that sometime they call Messiah the son of Ephraim, sometime of Joseph; for he shall come out of the Tribe of Ephraim, and shall be Captaine of all the ten Tribes, who gave their name to Ephraim, because that their first King Jeroboam was of that Tribe.[83]

Indeed, a decade later in New England, Increase Mather gave a series of sermons on the prophecies about the Jews and Israelites in which he cited Menasseh's *Hope of Israel* and spoke about Jeroboam and the 10 tribes.[84] All of this points to one conclusion: Jeroboam, Giorobaon or Giorbaon was a fictitious person based on a biblical character, which means that the Bible was informing a rumour that was perceived throughout 1665 and 1666 as a current event.

Although Giorbaon was mentioned in many of the Dutch sources, the German *Wahrhafftes Conterfey* and the *Warhaffte Abbildung Josuae Helcams* (1666) provided a competing army commander named Iosua Helcam. In particular, the *Wahrhafftes Conterfey* tied the sack of Mecca to the Sabbatian movement, noting that the prophet Nathan Levi had anointed a certain young man named 'Sobeza' (read Sabbatai) as king and renamed him 'Iosua Helcam'.[85] According to the pamphlet, Iosua Helcam was the supreme leader of the army of the Lost Tribes in Arabia, and his exploits were plentiful.

While the name 'Giorbaon' came from the Jews and was derived from a story in the shared Judeo-Christian biblical books, 'Iosua Helcam' came from a German broadsheet of 1642 and was a pure Christian construction. 'Iosua' is related to 'Jesus' and 'Helcam' in Hebrew means 'God rose' (*hel* or *el* for God and *cam* or *qam* for rose).[86] The different versions of the rumour were taking on contrasting forms based on their routings across Europe.

It was the multitude of Dutch sources, be they pamphlets, newsbooks or gazettes, that kept the rumoured sack of Mecca in the news and convinced readers like Serrarius of its veracity. The Dutch Republic's tolerant policy allowed cross-religious interactions and the promotion of millenarianism and messianism that gave rise to an environment in which the rumour found new life. Multiple versions of it proliferated to an extent not witnessed anywhere else.

The rumour had developed to such a degree by the time it reached Amsterdam that Serrarius told of a letter from Florence which claimed that:

> the Court of Florence was much amazed at the news of the Israelites, because they said that the Emperor had written to the great duke, that he was drawing his Army from Germany,

83 See ben Israel, *The Hope of Israel*, 33, 15, 24.
84 Increase Mather, *The Mystery of Israel's Salvation* (London, 1669), 3–4.
85 Scholem, *Sabbatai Sevi*, 354.
86 Scholem, *Sabbatai Sevi*, 557–8.

and had raised a great army, and was going to fight with this people: In Florence they say, there are about 50000 ... [87]

In Casale Monferrato, the Jews conflated Nathan's prophecy of the Lost Tribes only slaughtering Christians in the Ashkenazi lands of Germany and Poland with the rumoured siege of Mecca to claim that the ancient Hebrews were soon to leave the holy city of Islam and march against the Germans and Poles. In Amsterdam, supposed German diplomatic correspondence written to the Tuscan authorities reaffirmed this addition to the rumour, corroborating the Lost Tribes' intended invasion of Germany from the Germans themselves.

With all of these handwritten and published documents presenting the same narrative, Serrarius accepted the truthfulness of the story and repeated it in letters to his friends and associates. Citing what appears to be Raphael Supino in Livorno, Serrarius wrote to John Dury:

> I shall inform you of what seems to be incredible. The city of mecca, the seat of the Mohammedan superstition, is now besieged by a people calling themselves the children of Israel and saying that they were merely the vanguard of the army of their brethren who were following them. The news, which arrived here from Leghorn three weeks ago, was sent by a Jew who says that he heard it from a Jew who had come from Alexandria in Egypt.[88]

Serrarius was not a passive transmitter. He understood the rumour within the context of his previous millenarian activities with Samuel Hartlib, John Dury and Menasseh ben Israel. Utilising his past knowledge, Serrarius created an entire back-story for the rumour, and the more the back-story developed, the more inaccurate it became. First, he combined the account of the Lost Tribes sacking Mecca with information that originated in a widely circulated letter from a Christian in Morocco:

> The tidings of the 15 of *July*, concerning the March of our Brethren, the *Ten Tribes of Israel*, is now from several places confirmed to Us, all *Things* being so full of wonder, that for some few days we could scarcely believe, or give credit to it, from the City of *Sus*, otherwise called *Santa-Crew*. But now We have certain Information, that they are on the side of the Desert, and move from several places to the said Desert Goth of *Morocco*, being not far from *Cape de Ver*, but more within the Land. And they appear daily more and more in great Multitudes, having suddenly and unexpectedly manifested themselves, covering a vast *Tract*

---

[87] Peter Serrarius, *The Restauration of the Jews* (London, 1665), 5: Scholem, *Sabbatai Sevi*, 335–6, correctly argued that these letters were compiled by Serrarius.

[88] Peter Serrarius to John Dury, as translated and quoted in Scholem, *Sabbatai Sevi*, 344–6. Dury then spread this news onward. He wrote to a correspondent from Bern on 28 October 1665: 'Mr. Serrarius, in letters from Amsterdam of September 15 and October 1, tells marvellous news of the Ten Tribes of Israel ... They have already made their appearance at the borders of Arabia, conquered Mecca, where the tomb of Mohammed is, and other cities, and put to death all the inhabitants except the Jews'.

of Ground, and consisting of about eight thousand Companies or *Troops*, each of which containing from one hundred to a thousand Men.[89]

Although this account had nothing to do with the one about Mecca and came from the opposite corner of the Mediterranean world, Serrarius saw both as part of a single storyline.

Serrarius was not the only Christian to weave these two narratives together. As early as September 1665, an English correspondent asked Serrarius about a report that the Jews were encamped before Mecca, awaiting the arrival of the main body of the army that had appeared in Morocco.[90] The *Herstellinge van de Joden* printed in Amsterdam, the *Ydele Verwachtinge der Joden* (1669) written in Smyrna and *The Last Letters To the London-Merchants and Faithful Ministers* (1665) published in London all discussed these stories side by side too. The *Historis Verhael* (1665) even spoke of the leaders of the Sabbatian movement in combination with the army from Morocco: the 'Koning der Joden; Sabatha Sebi' (king of the Jews: Sabbatai Sevi) and his 'Propheet Nathan Levi' (prophet Nathan Levi) had gathered a large following of Jews 'In't Landt van Sus, zijn 8000 troupen ... In Barbaryen, in de Woestijne van Theophileta, zijn ongevaer hondert duysent Joden' (In the Land of Sus, there are 8000 troops ... In Barbary, in the desert of Theophileta, there are about 100,000 Jews).[91]

Serrarius had a broader outlook than most Christian readers and writers. He placed both rumours in a global context. In *The Restauration of the Jews* (1665), the Dutch scholar wrote:

> Those [Jews] in Arabia are of the same company with them that appear south of Morocco, and all of them seem to lye hid in the Inland Countrey of Africa, extending themselves over the vast Tract of Land comprehending all between the two Tropicks, almost as far as the Cape of Good Hope ... such as went from them are one half of the way to Meka; he thinketh they possessed the Arabians Countrey, and went out of Africa into America by the strait of the entry into the Red-Sea; but whether by Boat or Miracle, he knows not ...[92]

Serrarius then turned to the itinerary outlined by Menasseh ben Israel in *The Hope of Israel* (1650) to explain how the Lost Tribes travelled from Africa to the Americas, tying the rumoured sack of Mecca to a theory postulated over a decade before in

---

[89]   This was found in the Dutch *Historis Verhael van den Nieuwen Gemeynden Koning der Joden; Sabatha Sebi, als Mede sijn by Hebbende Propheet Nathan Levi* (1665) as well as in the English *The Last Letters To the London-Merchants and Faithful Ministers* (London, 1665). This report may or may not have been linked to the outbreak of the Sabbatian movement because stories of the re-emergence of the Lost Tribes appeared sporadically throughout the seventeenth century and, aside from the description of the leader of the Lost Tribes in a manner that combined features of Sabbatai and Nathan (he had Sabbatai's stature, obesity and complexion and Nathan's gift of discernment), there is nothing in it that suggests a connection between this account and the Sabbatian movement. For more, see Scholem, *Sabbatai Sevi*, 350.

[90]   Scholem, *Sabbatai Sevi*, 334.

[91]   ULL *Historis Verhael*, 5–6.

[92]   *The Restauration of the Jews*, 2–3.

which the 10 tribes of Israel were part of a single body spread across the world from the Americas to China.[93]

After describing what had happened, Serrarius told his audience what would happen. The Lost Tribes were currently in the first part of their two-stage religious evolution. First, they would turn away from idolatry to proper Judaism. Then they would convert to Christianity,[94] which would lead to a period of universal worship among all peoples of the world in 1672.[95]

Serrarius' choice of 1672 is significant because this was also the year that Nathan prophesied that Sabbatai would return from beyond the river Sambatyon with the Lost Tribes. Considering Serrarius' close friendships with Jews in Amsterdam who were aware of this prophecy, the Protestant scholar would most likely have known about it too, and it could have affected Serrarius' choice of 1672 as the year of coming bliss – a possible cross-religious influence.

Serrarius shared his ideas with numerous correspondents, and some of his letters were printed in English and Dutch pamphlets. The editors, however, perceived Serrarius' universalism as unacceptable for the English Christian audience. In a published letter that relayed the abovementioned story to Nathaniel Homes, Serrarius supposedly claimed that in 1672, 'there will be a full Communion and Restitution of the Apostolistic worship, And then will the Gospel be preached throughout the whole world'.[96] Since Serrarius' original letter spoke in more general terms of universal worship and peace, not 'Apostolistic Worship' or the 'Gospel', the printed letter to Homes demonstrates how Serrarius either expressed his universal religious beliefs in Christian terms to this audience or how Homes re-Christianised Serrarius' ideas before disseminating them among the larger English population.[97]

'Meccha-news' in the Royal Society

Petrus Serrarius' correspondents in England included Henry Oldenburg, the secretary of the Royal Society. Serrarius acted as Oldenburg's intermediary in his ongoing academic communication with Baruch Spinoza, and Serrarius supplied Oldenburg with European news relating to their common political, intellectual and religious concerns. For at least one of these reasons, Serrarius thought that Oldenburg would be interested in the sack of Mecca.

---

[93]   *The Restauration of the Jews*, 2–3.

[94]   Goldish, *The Sabbatean Prophets*, 157.

[95]   Peter Serrarius to John Dury, as translated and quoted in Scholem, *Sabbatai Sevi*, 344–6.

[96]   *The Last Letters To the London-Merchants and Faithful Ministers*, 3.

[97]   Serrarius' teleological vision of religious evolution resulting in universal worship is markedly similar to that of Maimonides, whose work had been published in Latin and was known by seventeenth-century Christian scholars. The students of Serrarius' friend Menasseh ben Israel were often responsible for translating Maimonides, and one wonders if Serrarius was influenced by Maimonides' ideas. For more on Christian interest in Maimonides, see Aaron Katchen, *Christian Hebraists and Dutch Rabbis: Seventeenth Century Apologetics and the Study of Maimonides' Mishneh Torah* (Cambridge, Mass.: Harvard University Press, 1984).

The earliest reference to the rumour in Oldenburg's correspondence occurred in July 1665, which was before the majority of the reports were printed in the news sources in Italy, the Dutch Republic and England. This means that Serrarius must have told Oldenburg about the attack on Mecca as soon as he heard about it, not waiting until he received the confirmation he desired in the Dutch gazettes. Oldenburg shared it at an equally fast rate, showing the movement of information through multiple correspondence networks from the Ottoman Empire to England before it was published in places much closer to the Levant.

Oldenburg destroyed almost all of the letters from Serrarius during this period, most likely due to fear of punishment for consorting with the enemy during the Anglo-Dutch conflict, which makes the exact nature and content of the connection impossible to know. Oldenburg only subtly indicated that Serrarius informed him about Mecca's destruction in his correspondence with Robert Boyle. Boyle, one of the key figures in the scientific revolution of the seventeenth century, engaged in profuse exchanges with Oldenburg in the 1650s and 1660s, focusing primarily on Boyle's scholarly work.[98]

Like the letters that he received from Serrarius, the ones that Oldenburg sent to Boyle often included the latest news. At the end of September 1665, Oldenburg wrote to Boyle, 'Yt ye French letters assure ye death of the K. of Spain; and ye Dutch ye Meccha-news. M. Serrarius tells me, yt ye transcribing of Mr. Borrels manuscript goes on a pace ... '.[99] A week later, Oldenburg continued, 'I cannot adde any more, yn to intimate in a word, yt the Meccha-news grows stronger ... I know not, Sr, whether I mentioned to you in my last Mr. Serrarius, his desire, about the disbursing of some mony for the transcribing of Mr. Borrels papers'.[100]

In Royal Society circles, the rumour gained broader oral dissemination in part because of scepticism towards it. Boyle originally replied to Oldenburg:

> You had reason to think I would looke upon your News about ye siege of Meccha as very strang& soe do's Dr Pocock to whom I imparted it, though he suspects it may be some other of the neigbouring people, who have formerly both attaqu'd & plunderd Mecca ... [101]

In trying to verify or develop his opinions, Boyle had sought advice from Edward Pococke, the professor of Hebrew and Arabic studies at Oxford who had been the chaplain of the English colonies in Aleppo and Istanbul. After his return home, Pococke maintained correspondents in Aleppo and was probably one of the few

---

[98]   Michael Hunter, *Boyle Between God and Science* (New Haven: Yale University Press, 2009), 192, 1–2.

[99]   Henry Oldenburg to Robert Boyle, 28 September 1665, as quoted in Hall and Hall, *The Correspondence of Henry Oldenburg II*, 534.

[100]   Henry Oldenburg to Robert Boyle, 5 October 1665, as quoted in Hall and Hall, *The Correspondence of Henry Oldenburg II*, 545.

[101]   Robert Boyle to Henry Oldenburg, 23 July 1665, as quoted in Hall and Hall, *The Correspondence of Henry Oldenburg II*, 444.

people in England who had sufficient knowledge of the Ottoman Empire to form an educated opinion on this matter.[102] But his remark that it was probably a neighbouring tribe that had plundered Mecca suggests that even he thought the city had been pillaged. Although Pococke and Boyle believed that the story was exaggerated and tried to determine the truth by analytically dissecting what they had been told based on their other knowledge, they did not question its overall authenticity. Like all the other approaches, this too was ineffective in deciphering what had really transpired in the Arabian Peninsula, which was most likely nothing at all.

Unlike most other sources, Oldenburg's extant correspondence did not supply any details. There was no mention of a leader's name or a description of the population that attacked Mecca, even though his information came from Amsterdam where more developed narratives circulated freely. Boyle's comment that it was probably a neighbouring tribe instead of the people Oldenburg suggested means that Oldenburg most likely provided Boyle with a fantastical account. Considering the versions known at the time as well as Oldenburg's connection to Serrarius, it was probably one that at least included the Jews if not the Lost Tribes. Coming from Serrarius, one would expect there to be an apocalyptic element to the story, but Oldenburg only framed the rumour in relation to the larger 'Christian–Islamic conflict'. Like the Italians, Oldenburg dwelt on the rumour's political, not eschatological, consequences: the Christians no longer needed to fear the 'Turks', he wrote, because they had 'work enough cut out for ym at Meccha'.[103]

The rumour of Mecca's destruction was so widespread by this point that Oldenburg himself was informed of it from multiple European correspondents. Such corroboration, however, failed to allay his doubts. He finished one of his letters by adding that his faith in the story of Mecca's destruction was growing weaker because it was not being confirmed with its original vigour.[104]

The English Press

Alongside his obligations to the Royal Society, Henry Oldenburg was employed by the undersecretary of state Joseph Williamson to work on the *Gazette*, England's preeminent news source. Because Oldenburg's job included supplying Williamson with newsworthy information, he could have facilitated the publishing of the rumoured sack of Mecca in the English press. The news items in the *Gazette*, however,

---

[102]   Pococke had approximately a hundred Jewish manuscripts, including three on the kabbalah. He continued to receive texts from the Ottoman Empire after his return through Levant Company merchants. Some of Pococke's letters from his friends in the Ottoman realms are found in MS Pococke 432 Fol. 11/12 in the Bodleian Library at Oxford.

[103]   Henry Oldenburg to Robert Boyle, 24 August 1665, as quoted in Hall and Hall, *The Correspondence of Henry Oldenburg II*, 481.

[104]   Henry Oldenburg to Robert Boyle, 24 August 1665, as quoted in Rupert Hall and Marie Boas Hall, eds, *The Correspondence of Henry Oldenburg: Volume III 1666–1667* (Madison, Milwaukee and London: The University of Wisconsin Press, 1966), 481.

did not match those in Oldenburg's correspondence. Moreover, Oldenburg was aware of the rumour in the summer of 1665, and the story did not appear in the *Gazette* until the middle of December, when it quoted a proclamation from the Dutch Jews from three months earlier:

> It is now about three month since the *Jews* gave out that near 600000 men were arrived at *Mecha*, professing themselves to be of the lost Tribes. Since which it is affirmed, that a new Prophet is arisen in *Jerusalem* ... This Prophet (say they) foretells the Restauration of the House of *Israel*, and to that purpose have invited a young man of the Tribe of *David*, called *Sabbatai Levi*, for their King, who was followed by thousands of people, and that he intended for *Constantinople*, to demand the Empire.[105]

The editor of the *Gazette* realised his error and issued a retraction within a week. It was not the Lost Tribes that conquered Mecca; recent letters from Aleppo to Amsterdam told a different story – the 'Bassa of Bissery' had revolted against the *Grand Signior*, which is supposed to be the ground of the story of the *Jewes*'.[106]

Since news editors in different countries often 'borrowed' stories from each other,[107] it should not be surprising that one of the Dutch gazettes printed an almost identical item as the English *Gazette* in the same week. Presented as a new report, the *Oprechte Haerlemse Saterdaegse Courant* noted that, according to new letters from Aleppo that arrived in Amsterdam, 'the Bassa of Bassary has risen up against the ruler', which 'must really be something, since the Jews are making such a noise about it'.[108]

The 'Bassa of Bissery' or 'Bassa of Bassary' mentioned in the English and Dutch gazettes most likely referred to the pasha of Basra. Since the Venetian avvisi discussed the sack of Mecca alongside a rebellion in Babylon, the news about the pasha of Basra in England and the Low Countries could have been a version of that uprising. Although originally identified as parallel events in Italy, the rebellion in Babylon was construed into the basis of the rumoured sack of Mecca in northern Europe. Months later, the *Gazette* itself reported that the 'Bassa of *Baslara* in the *Persian* Gulf, who has been so long in Rebellion, and so stoutly disputed it with four Baassaes, wherein he of *Babylon* fell, has at the last submitted'.[109] Could the length of time that the story of the pasha of Basra took to reach the English and Dutch presses have facilitated its

---

[105]   *Oxford Gazette*, 11 December 1665. The number 600,000 is a biblical figure that is found in other sources, including German pamphlets in 1508 and 1523. For more, see Andrew Gow, *The Red Jews: Antisemitism in an Apocalyptic Age 1200–1600* (Leiden: E.J. Brill, 1995), 136.

[106]   *Oxford Gazette*, 18 December 1665.

[107]   Brendan Dooley, ed., 'Introduction', *The Dissemination of News and the Emergence of Contemporaneity in Early Modern Europe* (Farnham: Ashgate, 2010), 11.

[108]   *Oprechte Haerlemse Saterdaegse Courant*, 19 December 1665, as quoted in van Wijk, 'The Rise and Fall of Shabbatai Zevi as Reflected in Contemporary Press Reports', 21, 21n. Meanwhile, the reports in the *Oprechte Haerlemse Courant* were disseminated across Europe to Russia, where several of them were printed in the Muscovite *Kuranty*. For more on Sabbatai in the Russian press, see Maier and Waugh, '"The Blowing of the Messiah's Trumpet"', 136–52.

[109]   *London Gazette*, 9 July 1666.

transformation into the supposed foundation of the rumoured attack on Mecca only to reach England at a much later date in its original form where it would then be seen as a separate event?

Despite the *Gazette*'s retraction, the rumour's ability to mutate allowed it to resurface a month later in England's most reliable news source. In January 1666, the *Gazette* published multiple items about a Jewish–Arab force laying siege to Mecca, not making any connection to its previous article about the Lost Tribes attacking the very same city. It stated, 'the *Jews* and *Arabs* had destroyed the Tomb of *Mahomet* at *Mecca*' and had taken several places.[110] A Jew named Giorbaon in Aden had gained mass support through his oratory skills and, under his guidance, his followers killed their pasha, forced the garrison to submit and began calling Giorbaon their prophet. Within three months, they had conquered numerous cities, including Mecca and Medina, and now wrote to their Jewish brethren that they would soon be free from 'the slavery of the Turk'.[111]

The different versions of the rumour were printed in England in the reverse order that they were created. While the narrative of the revolt in Babylon appeared chronologically before both accounts of the sack of Mecca, the English press told of the Lost Tribes attack on Mecca then the rebellion in Babylon and finally the Jewish–Arab army's sack of Mecca. This occurred because the stories moved along contrasting networks, which affected their dissemination. The report of the Israelites conquering Mecca fit into Protestant millenarian beliefs so it would have been understood to be more important and shared with greater haste, explaining why one of the latest versions to appear was published first in the *Gazette*.

One of the *Gazette*'s last news items about the violence in the Levant included none of the previous parties. No longer the Lost Tribes or Giorbaon, now it was the Jewish messiah himself who was leading an extremely large and violent military campaign against the Ottomans:

> *Constantinople, Feb. 19.* We have no small apprehension of these frequent Intelligences we receive, all of them bigg with relations of great Tumults in *Palestine; Sabadai,* their pretended Prophet, growing every day more powerfull; insomuch, as we have reports, that he leads no less then a hundred thousand after him, and is very severe against all Turks killing all they meet with ... There arrived yesterday a Vessel from *Ragusa,* who tells, that the two Ambassadors, sent by their Governor, with the usual Presents to the Grand Signor, were returned thither; and that the Bassa of *Jerusalem* had sent an Envoye, who was upon his way hither, with an Account of the many, and great Insurrections of the Jews in those Parts.[112]

In spite of the constant changes to the plot, English millenarians would have been most concerned with the overall story in which the Jews were believed to be helping

---

[110] *Oxford Gazette*, 22 January 1666.
[111] *London Gazette*, 30 January 1666.
[112] *London Gazette*, 8 March 1666.

the Protestants destroy the Muslim danger and liberate the land of Jesus. Soon enough they would all convert to Christianity and the end would come.

Because the rumour surfaced over and over again in reliable sources such as the *Gazette*,[113] most people who read these accounts would have at least wondered if something had not happened at Mecca. Most, but not all people. The outspoken opponent of the Sabbatian movement, the Jewish rabbi Jacob Sasportas, refused to accept the reports of the Lost Tribes that he heard from his correspondent Raphael Supino in Livorno. Moreover, the mathematician Sir Robert Moray told Oldenburg, 'The stories concerning the Jewes you find in our Oxford Gazette fid.credit Judeus &c'.[114] In other words, 'Let a Jew believe them'.[115] Moray's blunt assertion provides a counterweight to the responses of Christians such as Serrarius. While Serrarius would not believe the news until he read it in a gazette, Moray did not believe it when he did.

The reports in the English gazettes were supplemented by pamphlets, including *The Last Letters to the London-Merchants and Faithful Ministers* (1665), *The Congregating of the Dispersed Jews* (1666), *God's Love to His People Israel* (1666) and *The Restauration of the Jews* (1665), which all stated that the Lost Tribes had sacked the Muslims' holiest city. Unlike the *Gazette*, these pamphlets focused on the eschatological ramifications of this event. It was proof that the final days of history were at hand and Jesus' return was imminent. The hopes of the editors were obvious from title pages that began, 'Lift up your Heads, this is the Wonderful Year! 26th February, 1666'.[116]

Many of these works relayed the well-known narrative penned by 'Rapheck Supi', or the Jewish scholar Raphael Supino in Livorno. By the time that Supino's correspondence, which contained one of the earliest written accounts of the Israelites' plundering of Mecca, reached London, the rumour had become so pervasive that the pamphlets which printed Supino's letter did so alongside many other letters from Italy and the Dutch Republic that confirmed its claims.[117] The tale had multiplied to such an extent that it was being used to corroborate itself.

Published one after another, each pamphlet validated the previous one. *A Correspondent to Benjamin Levi* told of how Nathan of Gaza had said that 'on the 20th of this month Sebat [part of January and February] Reuben as the first-born of Jacob, will be at Gaza: but before that time, two men would come to order how the Redemption of Israel should be effected'.[118] A short time later, *A New Letter Concerning the Jews* reported that the prophecy had just been fulfilled: 'our Jews

---

[113] According to O'Malley, the *Gazette* itself was regarded as a reliable source of information by its readership. See Thomas O'Malley, 'Religion and the Newspaper Press, 1660–1685: A Study of the *London Gazette*', in Michael Harris and Alan Lee, eds, *The Press in English Society from the Seventeenth to Nineteenth Centuries* (London and Toronto: Associated University Presses, 1986), 33.

[114] Robert Moray to Henry Oldenburg, 15 December 1665, as quoted in Hall and Hall, *The Correspondence of Henry Oldenburg II*, 642.

[115] Hall and Hall, *The Correspondence of Henry Oldenburg II*, 643n.

[116] *A New Letter Concerning the Jewes* as quoted in Cecil Roth, ed., *Anglo-Jewish Letters (1158–1917)* (London: The Soncino Press, 1938), 70–72.

[117] See Serrarius, *The Restauration of the Jews*, 6.

[118] *A Correspondent to Benjamin Levi* as quoted in Roth, *Anglo-Jewish Letters*, 72–4.

yesterday received from Alcaire, Livorno and Venice, so many letters, and of so great credit, that all of them publickly in their Synagogues do now believe, that the Tribes of Ruben, Gad and half of Manasseh are come to Gaza, as the Prophet Nathan foretold'.[119] Indeed, the rumour had infiltrated such a breadth of sources in so many different forms that it would have been difficult not to at least consider it possible that something unbelievable was happening.

The stories of the Lost Tribes reappearing, first in Arabia and then in Morocco, sparked more and more reports of sightings of the ancient Hebrews. A pamphlet entitled *A Brief Relation of Several Remarkable Passages of the Jevves* (1666) elaborated upon the latest miracle. According to a letter 'very lately sent into England by a worthy man to his good Friend in London', 10 venerable men entered the king of Persia's court in October 1665 and professed themselves to be sent from Israelites that had been preserved in 'the remote parts of Tartaria'.[120] The men told the king that God had sent a prophet to gather his people together and lead them back to the holy land. After a brief struggle, the king allowed his Jewish subjects to leave and, with 'an Angel for their Conductor', they made their 'Plain-Way through the Mountains and Rivers' back to Palestine.[121]

At the same time, in October 1665, a letter from Aberdeen was printed in London under the title *A New Letter from Aberdeen in Scotland, Sent to a Person of Quality* (1665). This pamphlet told of the arrival of a ship in the harbour of Aberdeen that had sails made of 'white branched Sattin' that bore the inscription 'THESE ARE OF THE TEN TRIBES OF ISRAEL'.[122] The passengers were all Jews who wore black and blue clothing, ate only rice and honey, and spoke broken Hebrew. These mysterious people carried a letter, written in high Dutch for their brethren in Amsterdam, which related the stories already known to the English. There were 1,600,000 Israelites in Arabia and another 60,000 had come into Europe. Their forces had slain great numbers of Ottomans. None had been able to stand up against them. The Christians, however, had no need to fear. The Lost Tribes 'give liberty of Conscience to all, except the Turks, endeavouring the utter Ruine and Extirpation of them'.[123]

The English editor of this text was well aware of the broad dissemination of these rumours. He ended, 'I suppose you have not been ignorant of the Letters from several parts, the noise of them being communicated to most parts of Christendom', and he assured his readers that there is 'not a tittle in it but what is truth'.[124]

---

[119] *A New Letter Concerning the Jewes* as quoted in Roth, *Anglo-Jewish Letters*, 70–72. This pamphlet appears to have been written by Serrarius.

[120] NYPL *KC 1666: Lira Marashalack, *A Brief Relation of Several Remarkable Passages of the Jevves* (London, 1666).

[121] NYPL *KC 1666: Lira Marashalack, *A Brief Relation of Several Remarkable Passages of the Jevves*.

[122] R.R., *A New Letter from Aberdeen in Scotland* as quoted in Roth, *Anglo-Jewish Letters*, 68.

[123] R.R., *A New Letter from Aberdeen in Scotland* as quoted in Roth, *Anglo-Jewish Letters*.

[124] R.R., *A New Letter from Aberdeen in Scotland* as quoted in Roth, *Anglo-Jewish Letters*. Even in a city as close as Aberdeen, there were no known witnesses. It was only an unnamed professor who spoke to them.

As the plethora of stories demonstrates, the Israelites were apparently seen throughout the world. Not only in Arabia and North Africa, they appeared in Persia and even in Scotland too. After all, if the ancient Hebrews were showing up across the globe, should they not also be present on the islands of the country in which the news of their return was most widely documented?

'That of the Jewes' in the Puritan Atlantic

From England, at least one of these narratives was moved across the Atlantic along Puritan channels. At the beginning of September 1665, John Davenport wrote to William Goodwin:

> If that of the Jewes be true wee may easily see what god is bringing about in the world even the greatest changes that have beene since the 1st coming of Christ. The witnesses that are now killed, shall arise shortly. Rome shall be ere long ruined. Christ will take vnto himself his kingdome, which hath been vsurped by Brutish men the vileness of whose spirits hath appeared in scattering the churches unto Christ in silencing the faithfull ministers, in imprisoning and banishing the innocent in corrupting Religion with Antechristian superstitions.[125]

These words were written in America a mere four months after Sabbatai had publicly announced that he was the messiah in the Middle East, at a time when the Sabbatian movement was only known about among a select few in England. Even though the rumours of the Lost Tribes preceded the more concrete news of the messiah and his prophet, how would Davenport in North America find out about them so quickly?

While there are no extant sources that prove transatlantic transmission, the tale was first spread among Protestants in letters from Serrarius to Oldenburg and Dury, who were men that Davenport had been in contact with for years. Davenport wrote to John Winthrop in 1659 that he 'received letters and books, and written papers from my ancient and honourable friends Mr. Hartlib, and Mr. Durie'.[126]

Davenport's sparse reference was significant news that pertained to his eschatological expectations and, considering his transatlantic connections and the stories about the Jews that were circulating among his correspondents, it seems that the Puritans in the American colonies received word of the return of the Israelites through

---

[125]   John Davenport to William Goodwin, 2 September 1665, as quoted in Isabel MacBeath Calder, ed., *Letters of John Davenport: Puritan Divine* (New Haven: Yale University Press, 1937), 257. The death and resurrection of the two witnesses is part of the anticipated eschatological sequence as outlined in Revelation. Increase Mather would elaborate upon this in a treatise introduced by John Davenport: 'For the Jews (as hath been shewed) shall be converted before the day of judgment. And Rome (though not the whole Antichristian state) will be destroyed before the coming of Christ; and the resurrection of the two witnesses will take place before the sounding of the 7th [trumpet]'. For more, see Mather, *The Mystery of Israel's Salvation*, 143.

[126]   John Davenport to John Winthrop Junior, 19 June 1659, as quoted in Calder, *Letters of John Davenport*, 142.

this channel before most Christians in Europe read about it in the pamphlets – a startling demonstration of the speed in which news spread through correspondence across the Mediterranean and Atlantic worlds before it was printed in publications much closer to the Levant.

Years later, Davenport confirmed his knowledge of the rumours of the Lost Tribes in the preface to Increase Mather's *The Mystery of Israel's Salvation* (1669). Writing from his study in New Haven on 18 September 1667, Davenport made the following remark about a series of sermons preached by Mather late in 1665:

> These Sermons being preached in a time when constant reports from sundry places and hands gave out to the world, that the Israelites were upon their journey towards Jerusalem, from sundry Forreign parts in great multitudes, and that they were carried with great signs and wonders by a high and mighty hand of extraordinary providence, to the astonishment of all that heard it, and that they had written to others of their Nation, in Europe and America, to encourage and invite them to hasten to them. This seemed to many godly and judicious to be a beginning of the accomplishment of that Prophesie concerning the noise and shaking, and coming together of those dry bones spoken of in Ezek. 37.7. [127]

A footnote in the last sentence of this text stated, 'The Authors opinion (as is to be seen in the following Tractate) was and is, that the late and present rumours about the Jews, will prove an eminent, if not an ultimate impletion of that Prophesie, Matth. 24.24'.[128] This comment by Davenport, written two years after the rumour had reached its peak, was the first to describe it as a 'rumour'. Yet even Davenport acknowledged that Mather expected such stories about the Jews to be fulfilled shortly.

Mather was obviously influenced by, what Davenport had termed, 'that of the Jewes'. It inspired both sermons and his text, *The Mystery of Israel's Salvation*, which anticipated an armageddon battle of 'the Turk and Pope against the Kingdom of Christ amongst Jews and Gentiles' in which 'Asia is like to be in a flame of War between Israelites and Turks, Europe between the followers of the Lamb, and the followers of the beast'.[129]

Like other Protestants with philo-Semitic and Judeocentric millenarian leanings, Mather believed that the contemporary Jews and ancient Hebrews would help the Protestants destroy both the Islamic Ottoman Empire and the Catholic Church. This would then be followed by the conversion of the Jews and the Lost Tribes to Christianity as well as their restoration to Israel, which would bring about the advent of Jesus' thousand-year reign on earth.[130]

---

[127]   Mather, *The Mystery of Israel's Salvation*, An Epistle to the Reader.
[128]   Mather, *The Mystery of Israel's Salvation*, An Epistle to the Reader.
[129]   Mather, *The Mystery of Israel's Salvation*, 36–7.
[130]   This Judeocentric Puritan millenarian vision was also found in the writings of other English and American Protestants, including Thomas Brightman. For more on these ideas, see Cogley, 'The Fall of the Ottoman Empire and the Restoration of Israel in the "Judeo-Centric" Strand of Puritan Millenarianism', 304–32.

Mather's series of sermons inspired by the stories of the Israelites provide another medium of information dissemination affected by the rumour; however, the rumour did not infiltrate sermons to the same extent that it did most other sources. Mather's *The Mystery of Israel's Salvation* totals 175 pages, and the rumour only surfaces in footnotes and in passing on one or two occasions. Alongside the aforementioned notations, the only other possible reference to the attack on Mecca is found in a rebuke issued by Mather at doubting Christians. In reference to 'the providence of God in the present stirrings which we hear are amongst the Israelites', he states: 'I pity those which do in a scornful tryumphing manner say, "All that you hear is nothing, but a company of Arabians making a tumult": The Lord pity such men, they little what spirit they are of'.[131]

This may be somewhat unexpected because apocalyptic ideas were predominantly distributed by means of the sermon in this period, but the rumours of the Lost Tribes mainly appeared in broadsheets, pamphlets and gazettes instead. Sermons, such as Mather's, served to provide Christian audiences with the knowledge that would make these rumours appear legitimate when they were disseminated in the news sources.

The interaction between sermons and news possibly explains why the rumour was understood in such contrasting ways among Catholics and Protestants. In the American colonies, England and the Dutch Republic, this period was one in which sermons on the imminent end of the world were popular at the same time that treatises on eschatological matters, such as those by Thomas Brightman and Joseph Mede, were frequently reprinted. Often inspired by events rooted in local or national histories, millenarianism was on the rise in these places and preachers spoke of prophecies about the coming end.

The same cannot be said for Italy. The lack of interest in the prophecies in the Apocalypse and subsequent lack of an apocalyptic climate among the Italian Catholics in this period may aid in explaining why the same rumours were not understood in an eschatological manner. Catholics were not inundated with apocalypticism in the same manner as the Protestants and therefore did not read the rumour in the same light.

Returning to the Americas, the accounts of the Israelites were spread around New England, where they came to the attention of the Puritan missionary John Eliot. While Eliot had withdrawn his endorsement of the Lost Tribes theory in the mid-1650s,[132] these narratives re-invigorated his interest in the study of the ancient Hebrews, and he would remain intrigued by the Lost Tribes for the rest of his life.[133] The effects of the stories on the Englishmen in the American colonies highlight the far-flung cross-religious impact of the rumour: beginning among Ottomans in the Middle East, it spawned more and more versions that transcended the Mediterranean

---

[131] Mather, *The Mystery of Israel's Salvation*, 85–6.

[132] Cogley, *John Eliot's Mission to the Indians before King Philip's War*, 96.

[133] Cogley, 'John Eliot and the Origins of the American Indians', 221–2; Katz, *Philo-Semitism and the Readmission of the Jews to England*, 157.

and Atlantic worlds, influencing the beliefs of Puritans as far away as the east coast of North America.[134]

Back to the Ottoman Empire

The rumour did not simply travel westward from its origin in the Ottoman Empire; as it gained momentum it was transmitted back to the Levant. In 1667, the Dutch chaplain in Smyrna, Thomas Coenen, decided to write a pamphlet about Sabbatai Sevi after he witnessed firsthand the rise and fall of the Jewish messianic movement. Although Coenen's text, *Ydele Verwachtinge der Joden Getoont in den Persoon van Sabethai Zevi* (1669), or 'The Idle Expectations of the Jews, Shown in the Person of Sabethai Zevi their Last So-Called Messiah', was centred on the Sabbatian movement, it contained references to the attack on Mecca and the sightings of the Lost Tribes in Africa. Coenen quoted letters from the resident of the duke of Savoy in Lisbon, a Christian merchant from Paris and a German nobleman that discussed the Jewish army in Africa, the Israelites' plundering of Mecca and a description of Sabbatai Sevi respectively.[135] The manner in which Coenen mixed letters not only from different people and places, but also the different storylines within the letters, shows that he had access to a variety of European correspondence and too saw connections between the Jewish force in Africa, Mecca's destruction and the Sabbatian movement from his vantage point in the Empire.

The Red Jews in Germany: A Counterweight

While Protestant millenarians in England, the Dutch Republic and the American colonies looked forward to the return of the Lost Tribes, the situation in Germany was much different. Instead of seeing the Israelites as a force that would aid Christians in their battle against the Ottoman Muslims, German writers understood these rumours in relation to another tradition – that of the Red Jews.

In the twelfth century, Godfrey of Viterbo stated that Alexander the Great walled up both the Lost Tribes and Gog and Magog. The latter was an evil people spoken of in the Apocalypse who were expected to return near the end of time to fight against God's chosen people. By the second half of the thirteenth century, German scholars had conflated the Lost Tribes with Gog and Magog, creating a new population which came to be described in the German-speaking world as the Red Jews.[136] In the *Der*

---

[134]   Despite proof of dissemination to at least three English Christians in North America, there is no evidence that the rumour was known about in the Dutch American colonies, which had a long history of Jewish settlements as well as numerous networks connecting them to the Low Countries where these stories were shared broadly – an unexpected disjuncture in transmission.

[135]   Coenen, *Ydele Verwachtinge der Joden Getoont in den Persoon van Sabethai Zevi*, 131–2.

[136]   For more on the Red Jews, see Gow, *The Red Jews* and Rebekka Voss, 'Entangled Stories: The Red Jews in Premodern Yiddish and German Apocalyptic Lore', *Association for Jewish Studies Review* Vol. 36, No. 1 (2012), 1–41.

Jungere Titurel (1270), for example, the author claims to have encountered the savage and powerful Red Jews in Asia, where they were enclosed by a high mountain on one side and a river filled with stones on the other.[137]

After the Reformation, the Catholic position, as asserted by the Dominican Tomaso Malvenda and others, was that the antichrist would be a circumcised descendant of the Jewish tribe of Dan who would rebuild the temple in Jerusalem, re-establish the Jewish state and be accepted by the Jews as the messiah.[138] He would work false miracles, have false prophets and conquer Egypt, Libya and Ethiopia.[139]

Both the story of the antichrist and the legend of the Red Jews were popular and influential in early modern Germany, and scholars assigned the Red Jews to the ranks of the antichrist; one day they would break out from their prison and follow him. In the sixteenth and early seventeenth century, the Red Jews were described negatively in various German pamphlets as an army that would free the Jews from their servitude under Christians and liberate Jerusalem by sword.[140] Christians in Germany were 'terrified of this horrific people', who were expected to come forth in the last day to wreak havoc as the forces of the antichrist.[141]

Although reports of the Red Jews had disappeared by the end of the Thirty Years' War, this legend had a lasting effect: it shaped the manner in which the Sabbatian movement was portrayed in the German lands. During the height of the messianic excitement in 1666, a tale from Augsburg in 1642 resurfaced in Nuremberg and told of the coming of the Jewish messiah whose plan was to conquer the holy land with his military force. Unlike among the other Protestants, the advent of the messiah was not advertised in Germany as an imminent Christian triumph. It was the birth of the antichrist. This Jew had a neck that was 'thick-set, his head pointed, his face like that of a Turk, his brow wrinkled, his eyes terrible, his ears long, his penis large, and his teeth sharp'.[142] He was even said to be able to raise and equip an army of 50,000 soldiers from all over the world.

The German view provides a counterweight to the generally positive reception of these rumours by Christians. Even the term 'Red Jews' is manifestly opposed to that of Serrarius and others who thought these mysterious Jews were the 'white people whom

---

[137]   Voss, 'Entangled Stories', 4.

[138]   Howard Hotson, 'Anti-Semitism, Philo-Semitism, Apocalypticism and Millenarianism in Early Modern Europe: A Case Study and Some Methodological Reflections', in Alister Chapman, John Coffey and Brad Gregory, eds, *Seeing Things Their Way: Intellectual History and the Return of Religion* (Notre Dame: University of Notre Dame Press, 2009), 104, 102.

[139]   Richard Kenneth Emmerson, *Antichrist in the Middle Ages: A Study of Medieval Apocalypticism, Art, and Literature* (Manchester: Manchester University Press, 1981), 214–15.

[140]   Gow, *The Red Jews*, 78, 128.

[141]   Voss, 'Entangled Stories', 5.

[142]   Scholem, *Sabbatai Sevi*, 155. In numerous stories spread among Christians, the Jews' sexual character is highlighted. Some legends about Prester John state that he had to castrate the young Jewish slaves he captured because of their highly sexual nature. There is also a Dutch newsheet entitled *Cort Verhael van de Wonderen des Antichrist, Gebooren 1641, den 24 December, by Babylon ... Extract uyt Presburgh den 24 Junij 1642* (1643) that contains a very similar report.

the Inhabitants of Guiny use to speak'.[143] The colour red, after all, 'often held strong negative connotations in medieval and early modern Europe': it was equated with 'maliciousness and deceitfulness, dangerousness and ferocity' and used to 'stigmatize the enemies of Christ'.[144]

Death

Immediate unequivocal disproof from an authoritative source is the most effective way to quell a rumour.[145] But where was this to come from when the story of Mecca's destruction was found in such a wide variety of documents, and Mecca was one of the most difficult cities in Asia for Europeans to visit? The great distance between Mecca and northern Europe coupled with the lack of readily available communication to the Muslims' holiest city made disproving this tale incredibly difficult at best.

European Christians did not even have basic knowledge of Mecca. Both Protestant and Catholic writers incorrectly believed that Muhammad was buried there and not at Medina. Certain Ottoman Jews too lacked accurate information regarding the political situation in the Arabian Peninsula. The Jew from Alexandria who first shared the rumour with Raphael Supino in Livorno spoke of a 'king of Arabia', a non-existent figure who fought against a non-existent Jewish military corps.

Indeed, Arabia was so far removed from the regular networks of communication that the Muslim authorities in Yemen did not discover the first signs of the Jewish messianic movement until December 1666.[146] This was a full year after John Davenport in North America knew of 'that of the Jews', which means that the news travelled across both the Atlantic and Mediterranean worlds before it reached the more remote parts of the Ottoman state. The infrastructure of the early modern world allowed for the vast dissemination of information across two massive bodies of water before it went across a single desert. The implication of this was that the rumour could spread freely across much of the Abrahamic world because of the inability to check its veracity.

The cause of the rumour's rebirth, ironically, was also the root of its demise. The ubiquitous reports of the plundering of Mecca were eventually displaced by stories about Sabbatai, Nathan and the Jewish restoration. People grew weary of the news of Mecca's destruction; it ceased to be noteworthy and was replaced by more detailed,

---

[143] Serrarius, *The Restauration of the Jews*, 2. The Lost Tribes were often described as white. Twenty years earlier, Menasseh ben Israel, *The Hope of Israel*, 17, had thought he found the Lost Tribes after a Dutch mariner told him he had met a group of white, bearded, rich men in America.

[144] Voss, 'Entangled Stories', 15. For more on the possible meanings of the colour red, see this article.

[145] Ralph Rosnow and Gary Alan Fine, *Rumour and Gossip: The Social Psychology of Hearsay* (New York: Elsevier, 1976), 108.

[146] P.S. van Koningsveld, J. Sadan and Q. Al-Samarrai, *Yemenite Authorities and Jewish Messianism: Ahmad ibn Nasir al-Zaydi's Account of the Sabbathian Movement in Seventeenth Century Yemen and its Aftermath* (Leiden: Documentatiebureau Islam-Christendom, 1990), 80. For more on the Sabbatian movement's reception in Yemen, see this text.

more exciting and more pertinent accounts of the Sabbatian movement. If rumours continue until the underlying collective needs are fulfilled, or at least filled with something else,[147] then the growing messianic movement did just that. It pushed the sack of Mecca to the periphery and eventually out of public discourse.

The rumour, however, would not die that easily. Threads of it lingered on long after Sabbatai's conversion to Islam. Six years after the apostasy, the remaining Sabbatian believers who had gone underground still anticipated the fulfilment of Nathan's prophecy about the retribution against the Christians for their persecution of the Jews. In 1672, the year that Nathan predicted great events would happen, the Ottomans were victorious over the Polish army. The Sabbatians, or now Donmeh, attached special significance to this battle. According to them, 'the Turkish Sultan sent him [Sabbatai] at the head of 200,000 men to go to war against Poland in accordance with Nathan Ashkenazi's prophecy in Gaza that he would revenge the martyrs of Poland'.[148] Thus, parts of the rumour were adapted to the changing political circumstances in order to meet the needs of those who refused to believe that the messianic moment had ended. The Lost Tribes may not have returned, Mecca may not have been destroyed, but Sabbatai was still in command of a massive army. And this time, he was attacking the Christians, not the Muslims.

In 1709, over 40 years after the Venetian avviso first reported that Mecca had been sacked, the *Annals of the Universe* (1709) summarised the history of the rumour: 'the Jews reported at all places, that near 600,000 men were arriv'd at Mecha, professing themselves to be of [the] ten tribes and a half that had been lost for so many ages, but the story was false'.[149] The *Annals of the Universe* was correct, the story was false. Yet it failed to add that the rumour was most broadly disseminated by Christians, not Jews.

While previous chapters have centred on theories and news about real people, this chapter has shown that misinformation was as pervasive and could also tie individuals together across national, religious and continental divides. The rumoured sack of Mecca was an apocalyptic projection that points to an expansive map of misinformation. If the diffusion of an intellectual construct is connected to its news value,[150] then this rumour was perceived as more important than many of the true stories and intellectual theories already discussed. The sack of Mecca was written about in at least five languages in a variety of early modern communities on multiple continents within a single year. It surfaced in oral, handwritten and printed sources among travellers, newspapermen, natural philosophers and politicians.[151] An event of

---

[147]   Rosnow and Fine, *Rumour and Gossip*, 108.

[148]   Paul Fenton, 'Shabbetay Sebi and His Muslim Contemporary Muhammed an-Niyazi', in David Blumenthal, ed., *Approaches to Judaism in Medieval Times III* (Atlanta: Scholars Press, 1988), 86. The letter from Nathan of Gaza to Raphael Joseph stated that Sabbatai Sevi would persuade the sultan to be his slave and let him use the sultan's army to fight the Christians responsible for the 1648 massacres. See Liebes, *Studies in Jewish Myth and Jewish Messianism*, 96.

[149]   Benite, *The Ten Lost Tribes*, 71.

[150]   Rosnow and Fine, *Rumour and Gossip*, 32.

[151]   Although the rumour was spread among such diverse populations, there was a notable absence of women either as transmitters or as part of the population that comprised the Lost Tribes.

political significance for some and a sign of the coming end to others, the rumour took on different meanings for different people.

Considering the various versions of the tale together complicates previous historical understandings of the Sabbatian movement. No other scholar has identified the rumour's origins, realised its scope or recognised its complex relationship with the Jewish messianic movement. Abraham Gross wrote that the stories of the Lost Tribes 'came after the appearance of a Messiah in order to supply an answer to the anticipated need for Jewish military forces associated with redemption'.[152] Zvi Ben-Dor Benite offered another hypothesis: behind the news of the Lost Tribes sacking Mecca was the history of Jewish tribes in seventh-century Arabia. In Muhammad's day, the city of Khaybar was 'the great Jewish centre in the north of the Hajez', and legends abounded that the men of Khaybar were great warriors who belonged to the tribes of Reuben, Gad and Manasseh.[153] According to Benite, the persistent story 'of the Hijazi ten tribes' was at the basis of the rumours of the army of the Lost Tribes at the gates of Mecca in 1665 and 1666.[154] Matt Goldish, on the other hand, repeated and reaffirmed Gershom Scholem's opinion that it is 'by no means inconceivable' that reports of the Israelites spontaneously appeared with or without the input of rumours about Sabbatai.[155] None of these historians, however, was aware of the news items in the Italian avvisi, which means that none of them realised the original version only mentioned Arabs attacking Mecca and had nothing to do with the Jews.

It has recently been suggested that medieval Jews and Christians sometimes shared apocalyptic beliefs but understood them in inverse manners.[156] While this appears true in relation to the German understanding of the rumour, it was not the case for most Dutch and English Protestant millenarians who, like the Jews, believed that the return of the Lost Tribes signalled their redemption. If anything, this Judeo-Christian hope was inversely understood by Muslims as a catastrophe; however, it was a catastrophe anticipated in certain Islamic eschatological sequences. Thus, the re-emergence of the Israelites and their sacking of Mecca was a rumour that was thought to be much more than that, and it had the potential to appeal to Jewish, Christian and Muslim apocalyptic expectations.[157]

---

[152] Gross, 'The Expulsion and the Search for the Ten Tribes', 143.

[153] Benite, *The Ten Lost Tribes*, 75.

[154] Benite, *The Ten Lost Tribes*, 71.

[155] Goldish, *The Sabbatean Prophets*, 152.

[156] The Christian antichrist, for instance, could be seen as the Jewish messiah. Furthermore, both Christians and Jews used the story of Prester John and the Lost Tribes to their advantage in cross-religious discourse. For more, see Perry, 'The Imaginary War Between Prester John and Eldad the Danite and its Real Implications', 1–24.

[157] Filiu, *Apocalypse in Islam*, 47, has claimed that early modern Christian apocalyptic tradition inhabited a parallel world to its Muslim counterpart, sharing only a parallel register of images; 'there was no exchange, either positive or negative, between them'. The rumoured sack of Mecca at least partially calls this into question.

Chapter 4
A Jewish Messiah among Christians: The Evolution of European Perceptions of Sabbatai Sevi (1665–1666)

The rumour of the Lost Tribes' sacking of Mecca was spread alongside news of Sabbatai Sevi's messiahship. Both sets of narratives travelled from the Levant to the European Jewries through correspondence and word of mouth, moving along trans-Mediterranean Jewish familial and mercantile networks. Most of the letters about Sabbatai first entered Europe through the Italian port cities of Livorno and Venice, where many Jews, including scholars, accepted him as their messiah. Almost all of the Sephardic families in Europe had relatives in the Ottoman Empire who kept them informed of the latest events, so many of them were aware of the messianic outbreak even before Nathan's letters announcing the appearance of the messiah arrived in their communities.

When the correspondence about Sabbatai was passed along from Italy to Amsterdam, 'people were almost crushed in their eagerness to hear the most recent news'.[1] In the Ottoman Empire, the Sabbatian movement was tantamount to rebellion; in Catholic Italy, there was the constant danger of provoking the anger of the church or the state; but in the Dutch Republic, there was no need for dissimulation. The Jews' 'almost unanimous shouts of triumph were audible far and wide'.[2]

In Amsterdam, the continued arrival of letters from the Levant fanned the flames of the Jews' devotion, and the Dutch Sephardim tried to work out their restoration to the holy land in a practical manner. Due to the ongoing Anglo-Dutch war, Jean d'Yllan wrote to the king of England for a pass for a Dutch ship to sail peacefully to Jerusalem because 'God in his mercy has begun to gather in his scattered people and has raised up a prophet for us, therefore I and several of my Jewish brethren, together with fifty poor families desire to hire a ship to bring us to Jerusalem'.[3] In return, d'Yllan promised to 'pray for His Majesty's success'.[4]

---

[1] Scholem, *Sabbatai Sevi*, 532–3.

[2] Scholem, *Sabbatai Sevi*, 519.

[3] TNA SP 29/147/33: Jean d'Yllan to the king of England, 5 February 1666. While the original petition was written in French, it has been translated and summarised in the *Calendar of State Papers Domestic: Charles II 1665–1666*.

[4] TNA SP 29/147/33: Jean d'Yllan to the king of England, 5 February 1666.

Other Jews did not worry about such petitions; they set out towards Jerusalem themselves. The wealthy industrialist Abraham Pereira, the former patron of Menasseh ben Israel and a correspondent of Antonio de Montezinos, was in regular communication with the rabbi Meir Rofe, the head of a yeshiva in Hebron. With such connections, Pereira heard the tidings before almost anyone else and, upon learning that Rofe wished Pereira would join them to await the coming of the messianic age, Pereira left immediately for the holy land.[5]

In London, the small Jewish population heard about Sabbatai from Jews throughout Europe, including Raphael Supino in Livorno, who had been the companion of both Menasseh ben Israel and Jacob Sasportas on their journeys to the English capital. From northern Europe, the Sabbatian movement spread westward across the Atlantic to the West Indies.[6] With followers from the Middle East to the Americas, Sabbatianism became 'the largest and most significant messianic outburst in Jewish history'.[7]

It has been claimed that the 'false messiah had traumatic effects on Jews everywhere, but this passed unnoticed among their Christian neighbours'.[8] While the story of the Sabbatian movement among the Jews has been well documented, far less is known about how the Jewish messiah was understood and represented across Christendom. This chapter addresses the lacuna by providing well-known and newly uncovered letters, gazettes and pamphlets written in English, Italian, Dutch and Latin by Protestant and Catholic authors from the Ottoman Empire to the American colonies. An examination of the multitude of responses to the Jewish messianic outburst reveals that Sabbatai and his followers were a serious concern to many Christians for both spiritual and material reasons. In some cases, the Jewish movement led Christians to reconsider their own religious expectations and incorporate the Jewish messiah in a manner that blurred the boundaries between the Abrahamic faiths.

Studying the wide-ranging transmission of these letters and publications demonstrates that Christians understood and represented the Jewish messiah in contrasting ways due to their specific positions and beliefs. At the same time, comparing and connecting them show that these writings constitute an assortment of documents that travelled along different paths across the early modern Abrahamic world, often intersecting and changing in the process.

The Cross-religious Background to Sabbatianism

Sabbatai was not the first Jew to claim to be the messiah. For centuries, Jews had turned to the biblical books of Daniel and Isaiah with hopes for the coming of the messiah who would restore them to the land of Israel and usher in the age of peace.

---

5   Scholem, *Sabbatai Sevi*, 547.
6   Scholem, *Sabbatai Sevi*, 547, 549n.
7   Sharot, *Messianism, Mysticism, and Magic*, 86.
8   Monter, 'Religion and Cultural Exchange', 20.

Such expectations, especially when they surfaced in periods of persecution, led certain people to claim that they were the messiah.[9] Sabbatai was part of this long tradition that included many other messiahs, such as Jesus and his more immediate predecessor Solomon Molcho. In short, Sabbatai was first and foremost a product of this Jewish background. He was a Jew with rabbinical training whose messianic claims were based in his religious tradition.

There was, however, also the possible Christian influence on the Sabbatian movement that has already been discussed as well as an Islamic context that should be considered. The dispersion of the Sephardim brought the Spanish Jewry to the Ottoman Empire, where the rich tradition of Jewish messianism intersected with Islamic mysticism. The Jewish kabbalah developed primarily in Muslim lands after the Spanish expulsion,[10] and 'a certain factor within the Islamic context' shaped the development of Sabbatianism in both Muslim and Christian lands.[11] Although Nathan of Gaza's theology took on a larger Christian dimension as time progressed, he also drew upon a Jewish legacy of self-induced visions, which had intertwined with a corresponding Muslim tradition. Sufi techniques had interacted with the Abulafian method (especially in Palestine and Anatolia) over the centuries, and a 'striking connection' can be made between the Sufi ecstatic methods and the home of Luria, Vital and other sixteenth-century Safed kabbalists who were influential on Nathan: a Sufi prayer cave designed to promote ecstatic states was used in this period by the Safed kabbalists. Whether through the Abulafian literature that reflects an Islamic impact, the sixteenth-century kabbalists or direct local contact with Sufis, Nathan was influenced by Islamic mysticism.[12]

Christian Perspectives of the Jews' New Saviour

Literature penned by seventeenth-century Christians shows an awareness of the long history of Jewish messianism as well as the current expectations of their contemporary Jews. In the 1650s, Mary Cary wrote that the Jews 'are at this day earnestly expecting of, and waiting for the comming of the Messias'.[13] Around the same time, Thomas Goodwin noted that in 'that little book called, *The Voyage into the Levant*, he saith, That he had often occasion to converse with the Iews, whom he generally found to be in expectation of their Messiah'.[14]

Despite this, many Christians were shocked at the actual widespread outbreak of the messianic movement. The Sabbatian outburst affected the day-to-day business

[9] For more on Jewish messianism, see Scholem, *The Messianic Idea in Judaism*.

[10] Fenton, 'Shabbetay Sebi and His Muslim Contemporary Muhammed an-Niyazi', 81.

[11] Rapoport-Albert, *Women and the Messianic Heresy of Sabbatai Zevi*, 77. For more, see Scholem, *Studies and Texts Concerning the History of Sabbatianism and its Metamorpheses* (Hebrew), 100–120.

[12] Goldish, *The Sabbatean Prophets*, 34, 110, 61–2.

[13] Cary, *The Little Horns Doom and Downfall*, 139.

[14] Thomas Goodwin, *A Sermon of the Fifth Monarchy* (London, 1654), 23.

among the Jews that was of vital importance to European trade, so Christians became concerned with what was happening and sought to understand it. In Smyrna, the English merchants turned to their Jewish associates to find out more about Sabbatai and his adherents. A. Barnardiston, J. Adderley and N. Thurston 'had many discourses with the Jewes about this man [Sabbatai Sevi] from the Beginninge'.[15] In Amsterdam too, Christians noticed the change in the Jews' behaviour and started talking about it. After all, the Sephardim had significant investments in both the East India Company and the West India Company; their activities were of consequence to the Dutch economy. When the Sabbatian millionaire Abraham Pereira decided to leave for the holy land, a Christian author even reported the event in a newsletter:

> Abraham Perena, a rich Jew of this town parted on Monday last with his family for Jerusalem, after he had taken leave of our Magistrate, and acknowledged his thankfulness for the favour he and his Nation in their dispersion had received here etc. It's said he offered to sell a Countrey-house of his worth Three thousand pound Sterling, at much loss, and on this Condition, That the Buyer should not pay one farthing till he be convinced in his own Conscience, That the Jews have a King.[16]

Witnessing such whole-hearted devotion to this new cause, some Christians started to take the Jews' messiah seriously. But what were they supposed to make of him? The Lost Tribes of Israel had been incorporated into the Christian eschatological sequence and therefore the stories of their re-emergence in 1665 were understood as proof that Jesus was about to return. The reports of the Lost Tribes, however, were soon followed by news of the advent of a Jewish messiah, a person who had no role in any Christian apocalyptic scenario. The only messiah that was expected was Jesus, not an Ottoman Jew named Sabbatai. This was clearly a wrinkle in the plan. So how did Christians react?

Dismissing the 'False Messiah'

In every Christian community that received word of Sabbatai Sevi, there were those, possibly even the majority, who immediately dismissed Sabbatai as an impostor. All across Christendom, Catholics and Protestants saw him as another false messiah in a long list of Jewish messianic pretenders, and they mocked the Jews for being incredibly gullible and easily fooled.[17]

Such vehement reactions came from English intelligencers, diplomats and merchants stationed in London, Livorno and Smyrna respectively. A newsletter

---

[15]   TNA SP 97/18/211: A. Barnardiston, J. Adderley and N. Thurston to T. Dethick, 9 October 1666. Although the correspondence of these three English merchants has survived, not much else is known about them.

[16]   Untitled Christian newsletter from Amsterdam as quoted in Scholem, *Sabbatai Sevi*, 529–30.

[17]   Van der Wall, *De Mystieke Chiliast Petrus Serrarius (1600–1669) en Zijn Wereld*, 405.

written by Henry Muddiman described Sabbatai as 'a silly fellow, a baker's son'.[18] The *London Gazette* portrayed him as a 'pretended Messiah'[19] and one 'who pretends to be the Messias; for which [he is an] imposture'.[20] Meanwhile, the mercantile letters from Smyrna claimed that Sabbatai was 'an Imposture'.[21] For these Englishmen, the Sabbatian movement was merely a 'delusion'.[22]

Editors of newsbooks and gazettes in the Dutch Republic similarly referred to Sabbatai as the 'pretended' or 'so-called' Jewish messiah, using terms like 'Den gepretendeerde Joodse Messias'[23] and 'genaemde Joodse Messias'.[24] At the same time, Italian diplomats and avvisi editors frequently labelled him a 'falso messia'[25] and a 'pseudo Proffetta' too.[26]

If these Christians dismissed Sabbatai outright, then why did they waste their time reporting on his actions? The Sabbatian movement's negative impact on trade often inspired comments of disdain from the European merchants who depended upon the Jews as middlemen in their commerce. The Englishmen in Smyrna who witnessed the return of Sabbatai to his hometown as the messiah were painfully aware of the harmful effects of the messianic outbreak on their livelihoods: Jews closed their businesses to celebrate and spoke of leaving Smyrna permanently for Jerusalem. This caused such great concern for the English merchants that, in their regular letters to their business associate in Livorno, they wrote about the '[d]istraction amongst the Jewwes nation ... which is no small detriment to trade'.[27]

Although the merchants' primary concern was economic, they were curious to learn more about the theological foundations of the Jewish 'delusion' so they went to their minister, John Luke, in hopes that he could explain the Jews' beliefs. Luke was an educated pastor who told them that the Jews' claims relied upon God's promises in Zachariah – promises that Jesus had fulfilled when he entered the second temple. Because the temple was destroyed shortly afterward, the Jews would have to invent

---

[18] NA SP 29/151: Newsletter of Henry Muddiman to Edward Dyer, 15 March 1666.

[19] *London Gazette*, 15 October 1666. The English deputy in Livorno, Charles Chillingworth, similarly described him as 'the Jews pretended messia Sabata Sevi'. See TNA SP 97/18/233: although this is unsigned and undated, the handwriting appears to be that of Charles Chillingworth.

[20] *London Gazette*, 1 February 1666.

[21] TNA SP 97/18/156: A. Barnardiston, J. Adderley and N. Thurston to T. Dethick, 17 February 1666.

[22] *London Gazette*, 1 February 1666.

[23] ULL Castelyn, *Hollandtze Mercurius*, 134: August 1666 as well as ULL Castelyn, *Hollandtze Mercurius*, 33: March 1667.

[24] *Oprechte Haerlemse Saterdaegse Courant*, 4 September 1666, as quoted in van Wijk, 'Wachtend op. de Wolk naar Jeruzalem', 62.

[25] SA Segreteria di Stato, Avvisi 148: Turin, 17 June 1666.

[26] ASV Senato Dispacci Ambasciatori Constantinopoli F. 150, 19b–21: Giovan Battista Ballarin to the Venetian doge and senate, 18 March 1666.

[27] TNA SP 97/18/156: A. Barnardiston, J. Adderley and N. Thurston to T. Dethick, 17 February 1666.

new scriptures in order to prove the authenticity of Sabbatai's messiahship.[28] With this knowledge in hand, the merchants had both theological and mercantile reasons to portray the Sabbatian movement negatively in their continued updates to their partners in Europe.

Much of the news from the Ottoman Empire came to England through merchant letters, such as these, that were passed along to English intermediaries in Tuscany, who forwarded them to the secretary of state's office in London. Like the trade in goods, dispatches from the Levant to England often travelled via the port city of Livorno. In London, Joseph Williamson, the undersecretary of state and editor of England's official news source, would print the information from the letters in the *London Gazette*. Considering this, it should not be surprising that the items about Sabbatai in the English press were from Smyrna and repeated the merchants' concerns:

> *Smyrna, Jan. 18.* Our trade has been of late much obstructed in these parts, all the Jews being in a kinde of distraction upon the arrival of *Sabadai,* a Prophetical Jew, a *Smirnaite* born, lately come from *Jerusalem,* who in few days has preached such opinions into these people; that having perswaded them the number of years is accomplisht, all of them are big with expectation of their *Messias,* and the full restauration of the Jews.[29]

The Sabbatian movement's effect on trade similarly inspired negative portrayals in the Low Countries. While Dutch pamphlets published reports about Sabbatai from an early date because there was money to be made in doing so, the editors of the two prominent gazettes, the *Oprechte Haerlemse Saterdaegse Courant* and the *Oprechte Haerlemse Dingsdaegse Courant*, did not care about Sabbatai until his messianic following started to have an impact on the economy late in 1665. Then, between 19 December 1665 and 2 January 1667, the *Oprechte Haerlemse Courants* published 39 items relating to the Jewish messianic outbreak in 30 editions. Previously, they had rarely devoted more than five lines a year to Jewish subjects, so the large amount on Sabbatianism is astounding.[30]

These sources were renowned for their objective reporting, and the readership of a 'sensible newspaper such as the Haerlemse Courant' would have been interested in the Sabbatian movement for its mercantile effects.[31] Yet the gazettes' perspective of the Jewish messianic hopes was less than objective. Sabbatai's mission was a sad tragedy; the stories of his actions were mere fables. The gazettes only wished Sabbatai ill will.[32]

Jewish foreign trade was so important to the Dutch economy that multiple foreigners noted the concerns that were brought about by the Sabbatian believers'

---

[28]   TNA SP 97/18/156: A. Barnardiston, J. Adderley and N. Thurston to T. Dethick, 17 February 1666.

[29]   *London Gazette*, 12 March 1666.

[30]   Van Wijk, 'The Rise and Fall of Shabbatai Zevi as Reflected in Contemporary Press Reports', 24, 22.

[31]   Maier and Waugh, '"The Blowing of the Messiah's Trumpet"', 152.

[32]   Van Wijk, 'The Rise and Fall of Shabbatai Zevi as Reflected in Contemporary Press Reports', 23.

hoped-for restoration to the holy land. G. Willoughby wrote to Sir George Oxenden, the president of the East India Company at Surat, that the Jews were leaving Amsterdam for Jerusalem with expectations of their king, which would affect trade in the Dutch Republic.[33] Henry Oldenburg similarly told Lord William Brereton:

> The Holland letters continue their stories of the Jewes, and now tell us, yt they have appointed their Rendezvous at Jerusalem by ye first of Aprill, and yt the Jews at Amsterdam as well, as in other places, doe resigne their house, resolved to repaire for Palestina with the first conveniency. It may be, they will doe for so want of trade in Holland.[34]

While these men wrote about the Sabbatian movement for mercantile reasons, the Venetian diplomat in Istanbul, Giovan Battista Ballarin, thought Sabbatai was notable for the political implications of his following. No longer a religious impostor, Ballarin spoke of Sabbatai as if he were a leader of a rebellion against the Ottoman authorities: the Jewish messiah had forfeited his life as a 'auttor seditioso di comotione' (seditious author of a rebellion).[35] It was only Sabbatai's intelligence, sensible deportment and eloquent Arabic that saved his life since the vizier, a lover of languages, took a liking to him.[36]

Although Ballarin described Sabbatai negatively as a 'falso messia', he made numerous concessions. He stated, for instance, that the Jewish leader was 'di assai bella apparenza, ma di proffonda dottina' (of beautiful appearance, but [also] of great learning) who, after completing his severe penitential exercises, could make beams of light appear by his kabbalistic conjuration. A printed Italian avvisi from Turin agreed with Ballarin's assessment of Sabbatai's magical powers: it was 'con gl'effetti della sua eloquenza, e dell'Artemagica che possedeva, si era acquistato appresso i suoi il Titolo di Messia' (with the effects of his eloquence and the art of magic he possessed that he acquired for himself the title of messiah).[37]

Political and economic considerations were not the only reasons that Christians looked down on the Jewish messiah. Whether Catholic or Protestant, they also attacked Sabbatai in print in an attempt to defend their faith. Members of the

---

[33] G. Willoughby to George Oxenden, 5 March 1666, as quoted in McKeon, 'Sabbatai Sevi in England', 152.

[34] Henry Oldenburg to Lord Brereton, 16 January 1666, as quoted in Hall and Hall, *The Correspondence of Henry Oldenburg III*, 23.

[35] ASV Senato Dispacci Ambasciatori Constantinopoli F. 150, 19b–21: Giovan Battista Ballarin to the Venetian doge and senate, 18 March 1666.

[36] ASV Senato Dispacci Ambasciatori Constantinopoli F. 150, 19b–21: Giovan Battista Ballarin to the Venetian doge and senate, 18 March 1666. As Scholem, *Sabbatai Sevi*, 450, has noted, Sabbatai's knowledge of Arabic is highly unlikely since his Turkish was poor. This suggests that Ballarin was ill informed and was only passing along news he had heard secondhand. Europeans in the Empire, however, tended to speak highly of Sabbatai's linguistic abilities. An unnamed English correspondent in Smyrna, for example, wrote that Sabbatai spoke Turkish with the *cadi*. See TNA SP 98/6: unnamed author, 20 February 1666.

[37] See SA Segreteria di Stato, Avvisi, 148: Turin, 17 June 1666.

Jesuit mission in Istanbul sent letters about Sabbatai back to Italy, where they were distributed to 'uphold the Christian faith, and to ridicule the Jews as being an infidel'.[38] Meanwhile, at the height of the messianic excitement in Amsterdam, a Reformed Christian penned a treatise of 'sijn grondigh Bewijs, dat Iesus is de Ware messias, in den Paradyse belooft, ende van alle de H. Propheten voorseght, tot waerschouwinge der Ioden' (his thorough proof that Jesus is the true messiah, promised in paradise and proclaimed by all the prophets to the sight of the Jews).[39] Not content simply to prove that Jesus was the messiah, the author appended a short educational pamphlet against the Jews in order to counter their new beliefs.[40]

Even as far away as the American colonies, the Puritan minister Increase Mather responded to the Jewish messianic excitement by delving into the biblical scriptures, theological exegesis and political texts about the Middle East in order to better defend the Christian position. His diary is full of references to reading the apocalyptic works by John Cotton, Thomas Brightman and Joseph Mede as well as Henry Finch's *The Worlds Great Restauration or the Calling of the Jews*, 'Hornbeck contra Judaeos', 'Grovey of Turkish Empire' and 'Modern History of Turks, and Interest of Ctndome'.[41] All of his newly acquired knowledge came to the fore when he had '3 Jews with me whom I labored to convince that Messias is come'.[42]

Mather's journal entries prompt an intriguing question about transatlantic transmission. As with the rumour of the Israelites sacking Mecca, there is no definitive proof that the story of Sabbatai Sevi was known about among Christians in either New England or the New Netherlands. If these narratives were important for European Christians with extensive communication networks across the Atlantic, then why did they not come to the American colonies in a manner that we can track?

New Amsterdam was connected to the Dutch Republic through numerous mercantile, political and family networks. It was also home to a Jewish population that used their ties to the Dutch Sephardim to establish the first Jewish settlement in North America. When the Portuguese recaptured the colonies in Dutch Brazil in 1654, 23 Jewish refugees fled to New Amsterdam. While the governor of the New Netherlands, Peter Stuyvesant, tried to block every attempt by Jews to take part in community life, the intercession of the Dutch Jews led the West India Company to grant the Sephardim permission to settle in spite of the wishes of Stuyvesant.[43] Despite

---

[38]   Simonsohn, 'A Christian Report from Constantinople regarding Shabbethai Sevi ', 32.

[39]   *Oprechte Haerlemse Saterdaegse Courant*, 14 August 1666, as quoted in van Wijk, 'Wachtend op. de Wolk naar Jeruzalem', 62.

[40]   *Oprechte Haerlemse Saterdaegse Courant*, 14 August 1666, as quoted in van Wijk, 'Wachtend op. de Wolk naar Jeruzalem', 62.

[41]   See AAS Mather Family Papers Box 3, Folder 1: Increase Mather's diary.

[42]   AAS Mather Family Papers Box 3, Folder 1: Increase Mather's diary.

[43]   Noah Gelfand, 'A Transatlantic Approach to Understanding the Formation of a Jewish Community in New Netherland and New York', *New York History* Vol. 89, No. 4 (2008), 381. For more on Jewish settlement in North America, see this article.

such strong ties, there are no direct references to the rumours of the Israelites or the news about Sabbatai in any of the surviving sources.[44]

In New England, Mather's diary only provides circumstantial evidence. In his series of sermons published under the title *The Mystery of Israel's Salvation*, the Puritan pastor stated, 'Because the Jews received not the true Messias, therefore the vengeance of God gave them enough of false Christs'.[45] Mather then goes on to present a history of false Jewish messiahs from biblical days until the twelfth century. Considering that this text was published in London in 1669 and the preface was dated 1667, one wonders why Mather would not include the last and most popular 'false messiah' that was active within the last two years if he knew about him.

In sum, many Christians simply dismissed the Jewish messiah's claims. Some attacked the Sabbatian movement because of its economic or political implications, others in order to defend Christianity. Regardless of their reasons, they reached the same conclusion. Sabbatai was nothing more than an impostor. His followers were clearly deluded.

Toward Objective Reporting

Other Christians were not as eager to reject the Jewish messiah outright. The letters of many Europeans reflected a more 'ambivalent attitude'.[46] They wrote about the Sabbatian movement to inform or amuse their audience. An unnamed English correspondent in Smyrna, who witnessed the messianic movement firsthand, decided to write a multi-page report solely on Sabbatai in order to provide his associates in Tuscany and at home with a proper account of what was occurring in the Levant. This dispatch was a rare exception to the regular mercantile correspondence that dedicated no more than a paragraph or two to the Sabbatian movement in passing. This letter, which only survives among the papers of the Englishmen in Tuscany, presented what appears to be the closest known association between an Englishman and the Jewish messiah.[47]

The unnamed author did not even remember Sabbatai's name. He began, 'his name I have forgott, but he was borne in Smyrna, & two of his brothers live there, & his father was a broker'.[48] He knew that some Jews believed Sabbatai 'was a great prophet, others that he said he was Elias, others that he was the Messia', and even he accepted that it was certain Sabbatai 'was ever a man very much honored, & beloved

[44] Many thanks to Charles Gehring at the New Netherlands Project in the New York State Library for advice in trying to find such sources.

[45] Mather, *The Mystery of Israel's Salvation*, 85.

[46] Goldish, *The Sabbatean Prophets*, 159.

[47] According to Simonsohn, the French ambassador M. La Haye and some French gentlemen also visited Sabbatai in prison. For more on this interaction, see Simonsohn, 'A Christian Report from Constantinople regarding Shabbethai Sevi', 37.

[48] TNA SP 98/6: unnamed author, 20 February 1666.

for his upright conversation, fasting, & greate learning, having spent his time in study, from his youth'.[49]

Unlike the other Europeans, this Englishman was not content to take the Jewish informants at their word. He went to see Sabbatai himself.

> I found him in Mr. Penning brokers house; he satt according to the Custome of Turkey, upon a Sapria covered with Carpetts, leaning upon a cloath of Gold cushin, with a Fan in one hand, & a small glob in th' other being a very proper, comely man some small books lying before him, to all that came to doe him reverence he gave a little nodd with his head, he had likwise before him, all sortes of Fish, Flesh, and sweet meates, they that were admitted to the feast, satt all upon their knees, of each side of him, stood men in humble posture, houlding downe their heads, with one hand upon their brest, in y-e other, one houlding a glasse of wine, an other water, an other a silver pott, w-th sweete water, to cool his face or hands, an other a Pott of Incense, an other a Fann, but none offred to touch either meate or drink, for himself they sayeth but once in 3 dayes, all kept silence, except those that sang & plaied on all sortes of Turkish Instruments of Musick, all that attended him were richly clothed.[50]

Realising the cross-religious significance of Sabbatai's appearance in 1666 for Jews, Christians and Muslims, the author once again provided competing views of the Jewish messiah:

> He [Sabbatai] must be a very cunning fellow, that finding not only that its the greate expectation of all Christendome, something miraculous should be produced this yeare, drawne from scripture, but also the Jewes & Turkes themselves are possest of the same, hath taken this fit time to shew himself, when the world is disposd to receive any novelty, but others give their judgment, that he may be only a fit instrument in Gods hand to punish ye Jewes for their sins.[51]

The belief that members of other religions expected great changes in this period was echoed on the far side of the Atlantic Ocean in the prelude to Increase Mather's *The Mystery of Israel's Salvation*. Possibly embellishing the projections of Christian travellers based upon his own expectations, Mather wrote: 'not only Protestants, but Papists, Jews, Turks, Mahometans, and other Idolaters do expect some great Revolution of Affairs, as Travellers that have been amongst them do relate'.[52]

Returning to Smyrna, the unnamed correspondent wrote that it was others, not he, who judged Sabbatai as only fit to be God's instrument to punish the Jews. But the author's own Christian background was just below the surface, for he added: 'it was reported by severall, that some Jewes (though secretly for feare of ye rest) had

---

49   TNA SP 98/6: unnamed author, 20 February 1666.

50   TNA SP 98/6: unnamed author, 20 February 1666.

51   TNA SP 98/6: unnamed author, 20 February 1666. At another point, he added that 'so strangely are the people possessd with prodigious appearances in 1666'.

52   Mather, *The Mystery of Israel's Salvation*, To the Reader.

confessed, that he [Sabbatai] tould them, he had found by dilligent search, & study of ye Prophetts, that ye Messia was certenly come, & that was return'd to heaven againe'.[53] According to him, Sabbatai told the Jews that the messiah had already come and gone, a view that nicely coincides with the Christian position.

The Englishmen in Tuscany received many letters about the Jewish messiah from their associates in Smyrna, and some of them too did not use any disparaging language. The English deputy consul in Livorno, Charles Chillingworth, told Lord Arlington in London of the news that he heard from 'fresh letters this weeke from Constantinople' about the arrival of 'the Jewish Prophet'.[54] The English envoy in Florence, John Finch, similarly referred to Sabbatai as the 'Jews Messiah',[55] not a pretended messiah or an impostor.

Finch, however, took the news less seriously. He passed on 'the rumor of the Jews messiah ... desiring my Lord Arlington might have some divergent after his weighty affayrs are dispatchd'.[56] Finch thought that the story of Sabbatai was not news of significance; it was only important for its entertainment value. When the Sabbatian movement was still a topic of discussion among the Englishmen in Tuscany six months later, Finch continued to mention it in his regular dispatches to London. Yet his outlook had not changed. After presenting the latest political and military events in the Mediterranean world, he wrote:

> And since I have mentioned the Levant I cannot but acquaint your Lordship with the Table talke of the Jew's messiah: no less than 8000 people in one day go to visit Sabbatai in the castle where he is treated like a king, given a daily allowance, and is allowed to go abroad whenever he pleases.[57]

Unlike the merchants and diplomats in the Ottoman Empire, Finch considered the Sabbatian movement mere 'Table talke'. It was a popular topic of conversation, not a grave concern. Regardless, the information from Finch's letter was printed in the *London Gazette* a month later, when it reported:

> Fresh news is every day brought us of the great zeal of the *Jews* in the *Levant*, to the pretended *Messiah* who flock in such numbers to him, that in one day no less then 8000 strangers were

---

53   TNA SP 98/6: unnamed author, 20 February 1666.

54   TNA SP 98/6: Charles Chillingworth to Lord Arlington, 12 April 1666.

55   TNA SP 98/6: John Finch to Lord Arlington, 22 February 1666.

56   TNA SP 98/6: John Finch to Joseph Williamson, 12 February 1666. Ten days later, Finch reported to Lord Arlington that he had 'sent a relation of the Jews Messiah to Mr Williamson' in the same letter in which he noted that 'Yesterday by the convenience of my Lord of Winchelsea's chaplyn returnd the whole factory at my House heard the Common Prayer and a Sermon'. This suggests that the relation about Sabbatai possibly came to Tuscany with Winchelsea's chaplain. See TNA SP 98/6: John Finch to Lord Arlington, 22 February 1666.

57   TNA SP 98/7: John Finch to Lord Arlington, 18 September 1666. Finch too enclosed a copy of the notice that he received about Sabbatai, another source that has not survived.

in the Castle, where he is prisoner, to see him; that when he goes abroad (which is as oft as he pleases) he is always attended as a King, to the admiration of all sober men.[58]

Although Finch did not take the messianic movement seriously, the secretary of state was collecting accounts of the Jewish messiah – possibly because he thought that they were newsworthy. Sir William Temple wrote to the earl of Carlingford from Brussels in December of 1665:

> In my Lord Arlingtons last hee sent mee in payment of a fanatick relation I gave him of the Jews and their new Messias, and old prophecy w-ch hee read himself in an old manuscript at Oxford writ about Harry the 6th time, w-ch begins with the prediction of our loss of all wee had in France and God knows what it ends with but the Eagle having so much share in it, I thought it might entertain you at Vienna, for in these matters God knows I have not faith like a grain of mustard seed, though I censure none that has.[59]

Temple, a sceptic himself, did not 'censure' any who believed and forwarded the relation to the secretary of state without any critical remarks. The fact that Lord Arlington was collecting accounts of Sabbatai calls into question statements by historians who claim that Sabbatai was 'clearly not considered important either by the reporter himself [the English intelligencer Henry Muddiman] or by his superiors, among whom was Joseph Williamson. It seems certain that Muddiman's letter reflected Williamson's attitude and more generally the prevailing point of view in official circles'.[60]

Alongside the letters from the English diplomats, at least one Italian avviso did not describe Sabbatai as a false messiah as did the Turin avvisi, the English *Gazette* and the Dutch *Oprechte Haerlemse Courants*. Instead, a printed Italian avviso that appears to be from Genoa called Sabbatai the 'nuovo Principe degli Ebrei' (new Prince of the Jews).[61] It also told of how letters from the Egyptian Jewry to their European brethren urged the Jews to leave their homelands in order to go and bow down before their new prophet. Simply repeating the story and referring to Nathan of Gaza as the Jews' awaited prophet, the avvisi editor only labelled the news a 'curiosita'.[62]

As a curiosity, for entertainment or as information that was sought after or believed to be relevant, the reasons that these Christians wrote about the Sabbatian movement varied. Yet none of them openly condemned Sabbatai and his followers.

---

[58]   *London Gazette*, 15 October 1666.

[59]   BRBL Carlingford Papers Box 2, Folder 64, Series II: William Temple to Theobald Taaffe, 3 December 1665.

[60]   Zvi Loker, 'English Contemporary Opinions on the Sabbatean Movement', *Jewish Historical Studies* Vol. 29 (1982), 35–7.

[61]   SA Segreteria di Stato, Avvisi, 148: Genoa, 15 May 1666. Although this overly wordy term could possibly be ironic.

[62]   BNCF Codd Magliabechiani XXV, 743, 151b: Genoa, 9 January 1666.

Fear, Millenarianism and the Co-optation of Jewish Expectations

As Christians across Europe were informed of the Sabbatian movement in letters, pamphlets and gazettes, some became concerned with the Jewish messiah. Could the Jews be right? Had the messiah that they had spoken about for centuries actually come and, if so, what did this mean for Christians?

With a significant number of English publications about Sabbatai in circulation, the 'news of the messianic claimant and his followers made a greater impact on Christians in London than on Jews'.[63] Henry Oldenburg told Robert Boyle about two French letters which highlight Christian concerns over the Sabbatian movement. One was from the French ambassador in Istanbul, who wrote:

> Here there is great news about the King of the Jews, who is expected here soon; and it is said that the Sultan will be happy to yield him the Crown of Palestine. Most of the Jews have abandoned business, preparing themselves to go to Jerusalem. At first Mr. Legendre and I made fun of them, paying little heed to all this; but now appearances make us fear that all is not well.[64]

The other was from the French consul in Smyrna, who stated: 'Important news has reached here with the arrival of a King of the Jews in this city, a person of great consequence and wisdom; even the Turks hold him in esteem. Our nation lives in some fear. God grant that he will cause us no harm'.[65] Even the English *Gazette* remarked with apprehension about the advancing Jewish movement: '*Constantinople, Feb. 19.* We have no small apprehension of these frequent Intelligences we receive, all of them bigg with relations of great Tumults in *Palestine; Sabadai,* their pretended Prophet, growing every day more powerfull'.[66]

After hearing all of the reports about Sabbatai, Nathan and their Jewish followers, one Protestant author summarised the problematic implications of the stories:

> Being in a spirit of Judaism, in great power, led by a holy man, doing great miracles, and all things answering the description of the Messias, they may expect that it would be a testimony that the Christians and other people and Nations should be gathered in to the Jews, and not the Jews into Christ.[67]

While Sabbatai's claims were generally frowned upon in Catholic circles, the rise of philo-Semitism in some Protestant circles meant that certain English and Dutch

---

[63]   Endelman, *The Jews of Britain*, 32.

[64]   See Henry Oldenburg to Robert Boyle, 6 March 1666, as quoted in Hall and Hall, *The Correspondence of Henry Oldenburg III*, 49–51.

[65]   Henry Oldenburg to Robert Boyle, 6 March 1666, as quoted in Hall and Hall, *The Correspondence of Henry Oldenburg III*, 49–51.

[66]   *London Gazette*, 8 March 1666.

[67]   See Serrarius, *The Restauration of the Jews* (1665).

millenarians were ready to entertain news of the Jewish messiah with at least some enthusiasm because they made the connection between Jewish messianic hopes and their own expectations for the year 1666.[68] These Protestants anticipated the return of Jesus, the start of the millennium or the conversion of the Jews to occur. As John Sparrow noted in his diary in this year, 'Mr Jekyll said Mr More had no News of the Jewes, but they must certainly be called before the Beast be donne at 666'.[69]

Philo-Semitism gave way to Judeocentric millenarianism, which led a few Protestants to accept and co-opt the Jewish expectations, reframing them in terms of their Christian eschatological beliefs. Not sure what to make of the Jewish messiah, these Protestants were hopeful that the stories of the Jewish restoration would prove true. After all, the Christian apocalyptic sequence had long ago incorporated the Jews' return to the holy land, so this event could be understood as proof that Jesus' millennial kingdom was at hand. Upon hearing of the Jewish restoration, Oldenburg wrote, 'Here everyone spreads a rumor that the Jews having been dispersed for more than two thousand years are to return to their country. Few in this place believe it, but many wish for it'.[70] Around the same time, a report from Amsterdam that was printed in the English *Gazette* expressed bewilderment over the Christian interest in this news: 'It is strange, not onely the *Jews* here, but some hundreds that own the name of Christians among us, think themselves concerned in it'.[71]

Meanwhile, the Dutch *Haeghsche Post-Tydingen* featured a '[l]etter from a Catholic from Rome regarding the End of the World and the Jewish Messiah or King'.[72] Interpreting the letter in relation to his own beliefs, the editor stated that the Jews were 'speared like a harpooned Whale' and would 'embrace true Christianity/ which will be the end of their blindness: and the end of the World'.[73] He was certain that this must happen because the 'end of the World must as many say/ be nigh now ... that is, the year sixteen-hundred and sixty-six'.[74]

In the American colonies, some Puritans focused on the Jewish restoration that dovetailed with their beliefs and ignored the Jewish messiah. While none of their extant diaries or correspondence specifically referred to Sabbatai, many of them mentioned the hoped-for Jews' return to Jerusalem. John Davenport wrote from New Haven: 'If that of the Jewes be true wee may easily see what god is bringing about in

---

[68]   McKeon, 'Sabbatai Sevi in England', 161.

[69]   MS Rawlinson Essex 23: John Sparrow's diary, 1666. Mr. More is probably a reference to Henry More. An annotated edition of this journal will soon be released by Leigh Penman and Ariel Hessayon.

[70]   Henry Oldenburg to Baruch Spinoza, 8 December 1665, as translated and quoted in Hall and Hall, *The Correspondence of Henry Oldenburg III*, 637.

[71]   *Oxford Gazette*, 7 December 1665.

[72]   *Haeghsche Post-Tydingen*, 14 July 1665, as translated and quoted in van Wijk, 'The Rise and Fall of Shabbatai Zevi as Reflected in Contemporary Press Reports', 18: the front focused on the Anglo-Dutch war.

[73]   *Haeghsche Post-Tydingen*, 14 July 1665, as translated and quoted in van Wijk, 'The Rise and Fall of Shabbatai Zevi as Reflected in Contemporary Press Reports', 18.

[74]   *Haeghsche Post-Tydingen*, 14 July 1665, as translated and quoted in van Wijk, 'The Rise and Fall of Shabbatai Zevi as Reflected in Contemporary Press Reports', 18.

the world even the greatest changes that have beene since the 1st coming of Christ'.[75] Then, turning to the book of Revelation, he placed this news in terms of his Christian expectations. For him, it meant that '[t]he witnesses that are now killed, shall arise shortly'[76] – a necessary step in the apocalyptic sequence.

When Cotton Mather heard that the Jews were beginning to flock to Jerusalem with great signs and wonders, he thought that the vision of Ezekiel was on the verge of fulfilment: the dry bones of the whole house of Israel were being gathered to their home. Based on this report, he too concluded that the return of Jesus was imminent and wanted to hasten that day by converting the Jews to the 'true religion'.[77] Increase Mather also understood the stories of the Jews flocking to Jerusalem in a similar manner. It seemed to him, like it 'seemed to many godly and judicious [people] to be a beginning of that Prophesie [Ezekiel 37:7]'.[78] All of this, of course, was expected to bring about the conversion of the Jews and the millennial kingdom of Jesus.

Thus, the rise of philo-Semitism and Judeocentric millenarianism led certain Protestants to co-opt the Jewish beliefs, understanding them in terms of their own Christian eschatological expectations. They were either unaware or chose to ignore the messiah at the same time that they accepted the Jewish restoration, which they too perceived as necessary before the end would come.

Cross-religious Fusion

The co-optation of the Jewish messianic excitement by Protestants took its most extreme form among those Christians, like Petrus Serrarius, who mixed Jewish beliefs with their Christian ones to such an extent that they blurred the boundary between the two faiths. Serrarius, who was the first Christian to write about the Jewish messiah,[79] became the source of much of the Sabbatian news that reached the broader Dutch and English populations. In the Low Countries, his letters about the Jewish movement were printed in pamphlets such as *Herstelling van de Joden* (1665) and *Translaet uyt een Brief van Sale in Barbaryeen, in Dato den 6 Augusti 1665* (1665).[80] In England, they appeared in *The Jewes Message to their Brethren in Holland* (1665), *The Restauration of the Jewes* (1665), *The Last Letters To the London-Merchants and Faithful Ministers* (1665), *Several New Letters concerning the Jews* (1666), *A New Letter* (1666), *Gods Love to His People Israel* (1666) and *The Wonder of All Christendom* (1666).

[75] John Davenport to William Goodwin, 2 September 1665, as quoted in Calder, *Letters of John Davenport*, 257.

[76] John Davenport to William Goodwin, 2 September 1665, as quoted in Calder, *Letters of John Davenport*, 257.

[77] Marcus, *The Colonial American Jew*, 1140: see Mather, *The Mystery of Israel's Salvation*.

[78] Scholem, *Sabbatai Sevi*, 549.

[79] Van der Wall, 'The Amsterdam Millenarian Petrus Serrarius (1600–1669) and the Anglo-Dutch Circle of Philo-Judaists', 90.

[80] Van der Wall, *De Mystieke Chiliast Petrus Serrarius (1600–1669) en Zijn Wereld*, 415. The latter is reproduced in Adri Offenberg, 'Uit de Bibliotheca Rosenthaliana', *Studia Rosenthaliana* Vol. 29, No. 1 (1995), 91–9.

Like many millenarians, Serrarius was fascinated with the stories from the Levant because they reaffirmed his belief that the world was about to end. He too focused on the Jewish restoration to the holy land, which he thought was necessary before Jesus would return. Serrarius believed that God was using Sabbatai to accelerate the eschatological scheme by bringing about the much anticipated restoration of the Jews.[81]

Unlike most millenarians, Serrarius took this belief one step further. He argued that the Christians should support the Jews and join them rejoicing in the holy land. In fact, he planned to travel there himself.[82] Serrarius displayed an unrivalled sympathy for Jewish messianism: he followed the Sabbatian movement keenly, supported the messiah's adherents in Amsterdam and interpreted Sabbatai's actions as a prelude to the messianic age.[83]

As a Protestant scholar, how could Serrarius incorporate a Jewish messiah into his Christian millenarian beliefs? It was his friend, the Jewish rabbi Menasseh ben Israel, who had supplied Serrarius with the intellectual theory that allowed him to give the Jews' new saviour a role within his Protestant apocalyptic scenario. Years earlier, Menasseh told Serrarius about the two-messiah theory, a Jewish development in which the character of the messiah was split into a messiah of the house of David and a messiah of the house of Joseph.[84] This theory was not widely known until Isaac La Peyrere used it, in his *Du Rappel des Juifs* (1643), to solve the biggest problem in Jewish–Christian reconciliation, namely the Christian claim that the Jewish messiah had already come. Queen Christian of Sweden was so enchanted with La Peyrere's ideas that she kept a copy of his treatise in her library and showed it to Menasseh when he came to visit her.[85] Menasseh too was captivated with the two-messiah theory and introduced it to the Christian community in 1654. Then, in *Vindicae Judaeorum* (1656), the Jewish rabbi proposed the two-messiah theory as the solution to the conflict between Judaism and Christianity:

> For, as a most learned Christian of our time hath written, in a French book, which he calleth the Rappel of the Iewes (in which he makes the King of France to be their leader, when they shall return to their country,) the Iewes, saith he, shall be saved, for yet we expect a second coming of the same Messias; and the Iewes believe that the coming is the first, and not the second, and by that faith they shall be saved; for the difference consists onely in the circumstance of the time.[86]

---

[81]   Van der Wall, *De Mystieke Chiliast Petrus Serrarius (1600–1669) en Zijn Wereld*, 409–10.

[82]   Van der Wall, *De Mystieke Chiliast Petrus Serrarius (1600–1669) en Zijn Wereld*, 409–10.

[83]   Van der Wall, *De Mystieke Chiliast Petrus Serrarius (1600–1669) en Zijn Wereld*, 627. Serrarius provided his perspective of Sabbatai in the prologue to his *Verklaringe over des Propheten Jesaia Veertien Eerste Capittelen* (1666).

[84]   Scholem, *The Messianic Idea in Judaism*, 18.

[85]   For more on Menasseh ben Israel's relationship with Queen Christina of Sweden, see Katz, 'Menasseh ben Israel's Mission to Queen Christina of Sweden', 57–72.

[86]   Ben Israel, *Vindicae Judaeorum*, 18.

The two-messiah theory was not always looked upon positively by the Christian audience. In the following decade, Increase Mather would quote the Jewish rabbi's 'Spe Israelis' in his attack upon the two-messiah theory: 'The Jews have embraced an heretical fiction that there shall be two Messias's, one of the Tribe of Judah, and another of the Tribe of Joseph'.[87]

Unlike Mather, Serrarius embraced this theory as a means to accept Sabbatai's messiahship without diminishing his faith in Jesus, and he wove the two religious traditions together in a manner that led him down the path of universalism: Serrarius believed that the approaching end of history would be accompanied by a period of universal peace for all people. He was a keen follower of the Jewish messiah because he thought that both Jews and Christians were partially blind; the truth was to be found in synthesising the knowledge that each had appropriated. Indeed, the complementary nature of Christianity and Judaism was at the heart of Serrarius' millenarianism.[88]

Serrarius grounded his acceptance of both the Jewish restoration and Sabbatai's messiahship upon this theological foundation. Serrarius anticipated the restoration of all 12 tribes of Israel long before 1665, so the stories of the return of the Lost Tribes that emerged in this year, which were tied to the Jewish messianic movement, prepared Serrarius to recognise Sabbatai as the messiah.[89] Serrarius even attacked his opponents who called these reports fables because, according to him, they did not believe in the Lost Tribes and therefore would not accept the truth of the Jewish king and prophet either.[90] The Dutch scholar even compared his Christian brethren's rejection of Sabbatai to the Jews' rejection of Jesus.[91]

Serrarius' correspondent, John Dury, was similarly intrigued with the Jewish messiah and tried to determine how Sabbatai fit into the eschatological sequence. Compared with Serrarius' unrestrained belief in the Sabbatian movement as the beginning of the Christian millennium, Dury's support was more moderate: Sabbatai was only the king of the Jews for the Jewish state. God was rewarding the Jews by giving them their messiah first because the Christians were not sufficiently reformed.

---

[87] Mather, *The Mystery of Israel's Salvation*, 130.

[88] Ernestine van der Wall, 'Mystical Millenarianism in the Early Modern Dutch Republic', in John Laursen and Richard Popkin, eds, *Millenarianism and Messianism in Early Modern European Culture IV* (Dordrecht: Kluwer Academic Publishers, 2001), 41–3.

[89] Three years earlier, in 1662, Serrarius had written: 'As to the Time of the Conversion of the Jews, and the restoring of the Kingdom of Israel (of which all the Prophets are full) and by consequence the end of the terrible 4th Beast of Daniel, that is at hand'. He also stated: 'But that it may more clearly appear, that those things which we have mentioned concerning the restitution of the Jews, and the end of the last Monarchy, and have by this Conjunction of the Princely or chief Stars inferred that they are at the door, are not meer Astrological Prognosticks, but matters founded in the Word of God'. See Peter Serrarius, *An Awakening Warning to the Wofull World by a Voyce in Three Nations* (Amsterdam, 1662), 27, 36.

[90] For more on Serrarius' beliefs in the Lost Tribes and the Sabbatian movement, see 'Hoofstuck X: Sabbatai Sevi, Nathan van Gaza en de Terugkeer van de Tien Verloren Stammen' in van der Wall, *De Mystieke Chiliast Petrus Serrarius (1600–1669) en Zijn Wereld*, 399–464.

[91] Van der Wall, *De Mystieke Chiliast Petrus Serrarius (1600–1669) en Zijn Wereld*, 408–12.

For Dury, it was most important that Protestants supported the Jews and worked toward Jewish–Christian reconciliation.[92]

The Fifth Monarchist Thomas Chappell was an English millenarian who also 'seems to have accepted Sabbatai's divinity, and the stories of pillars of fire above his head'.[93] A correspondent of Serrarius, Chappell repeated many of the Dutch Protestant's words in a letter to his friend James Fitton. The Fifth Monarchist wrote that, although things look low, the prophet Nathan will soon arrive and perform miracles. Meanwhile, Sabbatai had assured the Jews that 'Redemption … is at hand', and 'the Lord will soon … accomplish his works'.[94] In this manner, he connected the Jewish beliefs regarding their redemption to the Christian expectation of the restoration of the Jews and their plan of salvation for humanity.

Although Christian intellectuals often tried to define and defend their beliefs, the 'distinct divisions that scholars and theologians, both Christian and Muslim drew between their respective faiths, in practice were much more flexible and overlapped in the minds of the masses'.[95] Such a perspective explains one of the most unbridled Christian responses to the Jewish messiah. During the height of the Sabbatian movement, an unnamed Christian girl in Smyrna apparently proclaimed in an ecstatic outburst that Sabbatai was the messiah. She later said that she did not remember what had happened but went on to confirm her faith in Sabbatai before a priest and other people. While it is not known what type of Christian she was, if this actually occurred, she serves as an example of a Christian who became a Sabbatian prophetess.[96]

The rise of the Jewish messianic movement therefore triggered a broad spectrum of Christian responses. Some simply dismissed Sabbatai outright. Others were more cautious and withheld judgement. They repeated the stories that they heard to inform or entertain their audience. At the same time, the growth of philo-Semitism and Judeocentric millenarianism led certain Protestants to co-opt the Jewish hopes. Most of these people only accepted the Jewish restoration to the holy land, which was part of the anticipated Christian apocalyptic scenario; however, a few of them fused the Jewish and Christian beliefs together to create new ones that blurred the boundary between Judaism and Christianity.

---

92   While Dury has been identified as a millenarian, a closer reading of his apocalyptic commentaries reveals that Dury's apocalyptic beliefs were more complex. For more, see Gibson, 'John Dury's Apocalyptic Thought: A Reassessment', 299–313.

93   Capp, *The Fifth Monarchy Men*, 214. A letter based on material from Serrarius read: 'The Jews have received a letter … from one at Smyrna who accompanied the King [Sabbatai] to Constantinople … he was at a palace with rabbi, the book of the Law … a pillar of fire … was seen hovering over the place where he was appearing in a Bed Chamber and they sath a confirmation from a vision he had and the man fell down dead'. See TNA SP 29/162/85: Thomas Chappell to James Fitton, 12 July 1666.

94   TNA SP 29/162/85: Thomas Chappell to James Fitton, 12 July 1666.

95   Dursteler, *Venetians in Constantinople*, 183.

96   Goldish, *The Sabbatean Prophets*, 159–60.

The Islamic *Dajjal* or Antichrist

Although some English and Dutch Protestants accepted the Jewish messiah's ability to perform miracles, none of them, not even his harshest critics, considered Sabbatai the antichrist – the evil person who is expected to have miraculous powers, gain power for a limited period of time and wage war on God's chosen people before Jesus returns and rules in his millennial kingdom.

Certain Muslims, however, did see Sabbatai as their apocalyptic enemy. In the Islamic eschatological sequence, Muslims anticipate the arrival of a figure similar to the Christian antichrist known as the dajjal (in Arabic). Considering the shared nature of the Abrahamic faiths, it should not be surprising that Muslims believe that the dajjal will have miraculous powers, raise an army of deceivers, gain dominion for a limited period of time that is either 40 days or 40 years and allow impurity and tyranny to rule the world. In some traditions, Muslims even anticipate the return of Jesus, who will fight and defeat the dajjal, ushering in a period of bliss in which Islam will be universally accepted by all people.[97]

The appearance of the dajjal is one of the proofs that the end times is upon humanity, and more than one Muslim believed that Sabbatai was the dajjal. In Yemen, Arabic writers represented Sabbatai and his supporters, whose movement expected the end of Islamic power, as the dajjal and his evil forces.[98] In Istanbul, the Bektashi leader Muhammad Niyazi was a friend of Sabbatai whose own Muslim followers believed that he was the messiah too. It 'seems obvious that his [Niyazi's] eschatological effervescence contributed in no small measure to similar frenzied reactions within the Jewish camp. Conversely, the appearance of a Jewish *daggal* [dajjal] and the excitement which attended it could only but have heightened this Muslim messianic fever'.[99] In short, the dovetailing Jewish and Islamic messianic movements coupled with the Islamic interpretation of Sabbatai as the dajjal created a reciprocal cross-religious influence that contributed to the growth of eschatological excitement among members of both faiths.

The Turning Point

Returning to Christendom, stories about the Lost Tribes and the Jews' messiah circulated in correspondences, pamphlets and gazettes across Europe, keeping Christians informed of the messianic excitement that had continued uninterrupted through Sabbatai's arrest and imprisonment. Anticipation was at its peak across the Jewish world when Sabbatai was finally called before the Ottoman divan, and one Jewish Sabbatian believer in London was willing to bet, with ten to one odds, that

---

[97]   For more on Islamic eschatological beliefs, see Cook, *Studies in Muslim Apocalyptic.*

[98]   Van Koningsveld, Sadan and Al-Samarrai, *Yemenite Authorities and Jewish Messianism*, 14.

[99]   Fenton, 'Shabbetay Sebi and His Muslim Contemporary Muhammed an-Niyazi', 87.

Sabbatai would be named by the Ottoman sultan and other political authorities as the 'King of the World and True Messiah'.[100]

When Sabbatai walked out of the interview as the new Muslim servant of the sultan, word of his apostasy spread rapidly. Despite their location far away from the Ottoman court, the English merchants in Smyrna knew about Sabbatai's conversion within three weeks.[101] In fact, multiple versions of the event had already proliferated by this time. Relaying the account that they had heard to their business associate in Livorno, the merchants wrote that everything began when one of Sabbatai's disciples, a 'polonese Jew', apostatised to Islam and claimed Sabbatai was a 'greate Impostor' who would lead the Jews into rebellion against the Ottoman authorities if the latter did not respond quickly. The Ottoman government convened and decided to massacre all of the Jews above the age of three to solve the problem. The heads of the Jewish community could not get the sentence revoked, but when Sabbatai was called before the divan and was asked to which religion he belonged, 'hee readily replyed that hee had searched into all but found none so right the truth as that of Mahomett which hee willingly entertain'd, relinquishing the law of Moses and by that means hath obtained the Jewes deliverance from the sword'.[102]

Transmitted along mercantile and political channels via the English colony in Tuscany, this narrative came to England in a slightly altered form and was printed in the *London Gazette*:

> [A] learned Jew from *Germany* (others say *Poland*) had seven or eight days conference with him [Sabbatai], who finding much vanity and weakness in his pretences and arguments, left him with much dissatisfaction and repairing to the Turkish Mosque turned Turk, declaring *Sabadai* to be an Impostor, and if care were not taken, he might draw the Jews into a Rebellion: of which, advice being sent to the Grand Signior, *Sabadai* was by order carried to *Adrianople*, where being examined by the *Chaymacham* and *Muftee*, he was by them sent to the Grand Signior in whose presence, He, with one of his Fellows, freely exchanged the Law of *Moses* for that of *Mahomet*, and was by the Grand Signior made a Cappegee Baffa, and called *Mahomet:* The same Letters tell us that the Grand Signior upon consideration of the great expectations and endeavors of the Jews to promote the interest of their new pretended King, has given command for cutting off all that Nation from seven years old and upwards, which upon their humiliation and intercession of the new Proselite was revoked ... [103]

Although it is not certain if genocide was actually threatened, this story bares a strong resemblance to that of Queen Esther in the Bible, who saved the Jewish nation from imminent slaughter through her intercession with the king. This similarity was noticed by the Englishmen in Smyrna, who noted that the Jews 'look on this deliverance not

---

[100]   Scholem, *Sabbatai Sevi*, 548. See the entry in Samuel Pepys' diary on 19 February 1666.
[101]   McKeon, 'Sabbatai Sevi in England', 155.
[102]   TNA SP 97/18/212: S. Pentlow, J. Foley and T. Laxton to T. Dethick, 29 September 1666.
[103]   *London Gazette*, 26 November 1666.

Inferior to that procured by Q[ueen] Ester,[104] and that the Jews believe it was 'Equal with that of Haman & Mordecai'.[105] The Esther narrative, a long-standing story of importance for Sabbatai's former Converso followers, was serving as a vital reference point for the Jewish Sabbatian believers.[106] It may have even informed the story of Sabbatai's conversion in a manner that justified and glorified his actions.

An English pamphlet by Serrarius titled *Gods Love to His People Israel* provided a competing account of the apostasy. Face to face with the Jewish messiah, the sultan asked him if he could perform miracles. When Sabbatai only answered, 'Sometimes', the sultan told him that he could either convert to Islam or be stripped naked and have arrows shot at him to determine if he was invincible. Not liking the latter option, Sabbatai chose to become a Muslim.[107]

A printed avviso in Torino similarly told of how the sultan wanted Sabbatai to perform a miracle; however, it claimed that the head of the Ottoman Empire warned the Jewish messiah that if he did not then he would be burnt to death. When Sabbatai saw them kindling a fire before him, he converted to Islam and the sultan assigned him a pension of 40 *soldi* a day to be his captain of the gate.[108]

One of the most unique descriptions of the apostasy that was distributed across Europe provided a version of the event that upheld Christian values. This narrative was written by a Dutch Protestant in Istanbul, but it first appeared in a printed text

---

[104]   TNA SP 97/18/214: S. Pentlow, J. Foley and T. Laxton to T. Dethick, 9 October 1666. This comparison found its way into the *London Gazette*, which reported that the Jews escaped severe punishment when Sabbatai 'freely exchanged the Law of *Moses* for that of *Mahomet*', which they 'look upon as a deliverance not inferior to that procured by Queen *Esther*'. See the *London Gazette*, 26 November 1666.

[105]   TNA SP 97/18/211: A. Barnardiston, J. Adderley and N. Thurston to T. Dethick, 9 October 1666.

[106]   The Bible tells the story of a Jewish maiden named Esther in Persia who was raised in exile by her cousin Mordecai. One day, the king wanted another wife and, after viewing all the virgins, he picked Esther. Meanwhile, Mordecai had made a mortal enemy of one of the king's counsellors, Haman, who devised a plan to have all the Jews killed in order to get his revenge. Haman duped the king into agreement, and Mordecai tore his clothes and begged Esther to make the king change his mind. But entering the king's presence uninvited was punishable by death so Esther and the Jews fasted for three days before she went before him. When she finally did, the king received her warmly and, in the end, she told him about her Jewish heritage and he revoked the planned genocide, ordering Haman to be hanged on the gallows he had built for Mordecai. Because Esther had dissimulated her beliefs but remained faithful to the religion of her ancestors at the king's court, the crypto-Jewish Conversos saw hope in the Esther narrative. Forced to live an 'Esther-like' experience where they too were subject to religious persecution and identity concealment, they drew parallels to the story of Esther, situating themselves in the Jewish tradition even though they were currently isolated from mainstream Jewish life. Furthermore, Esther saved the entire Jewish population through her actions, and the redemptive nature of the account suggested the possibility of the Conversos' own deliverance and return to Judaism. Conversos knew about Esther through the Catholic Vulgate, and 'Saint Esther' found her way into Converso theology – a name the Conversos gave her that itself highlights a cross-religious fusion. For more, see Miriam Bodian, *Hebrews of the Portuguese Nation: Conversos and Community in Early Modern Amsterdam* (Bloomington and Indianapolis: Indiana University Press, 1997), 15, 17; Miriam Bodian, '"Men of the Nation": The Shaping of Converso Identity in Early Modern Europe', *Past and Present* No. 143 (1994), 63.

[107]   Peter Serrarius, *Gods Love to His People Israel* (London, 1666).

[108]   SA Segreteria di Stato, Avvisi, 148: Turin, 16 December 1666.

from Rome entitled the *Lettera Mandata da Costantinopoli a Roma Intorno al Nuovo Messia de gli Ebrei* (1667). Told by the sultan that he must convert to Islam to remove the scandal he had created, Sabbatai replied that 'In tale frangente egli hebbe assai di compiacèva per quella nobil compagnia, avanti la quale egli si trovava per dire ad alta voce, ch'egli era Turco, e che voleva perseverare fino alla morte nella legge di Mahometto' (he was more pleased in his noble company, in front of which he found himself to say in a loud voice that he was a Turk, and that he wanted to persevere to the end of his life in the laws of Muhammad).[109] That is not enough, Sabbatai was told. He was ordered to 'confessi quì ad alta voce che non c'è altro Messia, che GIESU CHRISTO Figlio di MARIA' (confess here aloud that there is no other messiah than Jesus Christ Son of Mary) who came a long time ago.[110] Sabbatai responded that 'GIESU CHRISTO è il solo Messia' (Jesus Christ is the only messiah) and that it was mad and ignorant for the Jews to expect another.[111]

Protestant millenarians in Amsterdam and London knew about this story too. John Sparrow remarked in his diary, 'Serrarius has a Letter from a Protestant at Constantinople who wrote that Sabbathai was turned Turk having p[ro]fessedly declared yt Jesus Xt was the true messiah'.[112] Thus, the letter from a Dutch Protestant in Istanbul that was printed in the heart of the Catholic world and spread throughout northern Europe claimed that the Muslim Ottoman authorities forced the Jewish messiah to confess that Jesus was the true messiah when he converted to Islam – a narrative that blurred the religious boundaries between all three of the Abrahamic faiths.

Despite, and probably due to, the proliferation of so many versions of the apostasy, some European Jews refused to believe that Sabbatai had converted. According to a letter from Serrarius to Oldenburg, some Jews remained in denial for quite some time:

> As for the Jews their hope revives more and more. Those of Vienna having sent an Expres to Adrianopolis, do writ, that their Man doth affirm, to have spoken with Sabithai Sebi and found him, not turned Turck, but a Jew as ever in the same hope and expectation as before. Yea, from Smyrna by way of Marcelles we have, that at Constantinople the Jews return to their fasting and praying as before: and so doe some here likewise.[113]

Even the editor of the Dutch gazette thought that the Jews' refusal to believe the news was worth reporting. After telling of the messiah's apostasy, a news item in the

---

[109]   TNA SP 120/116: *Lettera Mandata da Costantinopoli a Roma Intorno al Nuovo Messia de gli Ebrei* (Rome, 1667).

[110]   TNA SP 120/116: *Lettera Mandata da Costantinopoli a Roma Intorno al Nuovo Messia de gli Ebrei* (Rome, 1667).

[111]   TNA SP 120/116: *Lettera Mandata da Costantinopoli a Roma Intorno al Nuovo Messia de gli Ebrei* (Rome, 1667).

[112]   MS Rawlinson Essex 23: John Sparrow's diary, 1666.

[113]   P[eter] S[errarius] to Henry Oldenburg, 5 July 1667, as quoted in Hall and Hall, *The Correspondence of Henry Oldenburg III*, 446–7.

Oprechte Haerlemse Saterdaegse Courant simply stated, 'De Joden gelooven noch niet de Tydinghe van haren Messias' (The Jews do not believe the tidings of their messiah).[114] Then, when confirmation of the apostasy came from multiple sources, first in a ship from Smyrna[115] followed by letters from Italy, the gazette eloquently summarised the conclusion of the messianic drama: 'soo wordt men van dromen wacker' (thus men wake from their dreams).[116]

'Greate Hopes'

As the Jewish world reeled in anger and dismay when news of the apostasy proved true, the Christians who had always considered Sabbatai a 'pretended messiah' found vindication. Almost all of the positive and objective reports about the messiah and his former followers disappeared; they were quickly replaced by letters, gazettes and pamphlets that lambasted Sabbatai as an impostor and his adherents as gullible and easily deceived.

The messiah's apostasy and subsequent failure of the Sabbatian movement were extremely 'detrimental to the Jewish side of the Jewish–Christian polemic'.[117] German, English, Italian, Latin, French, Dutch and Hebrew literary sources used the Sabbatian movement as an example to demonstrate Jewish blindness.[118]

Christian disdain, however, was accompanied by great hopes for the Jews. First and foremost, Christian writers hoped that the Jews would return to the business and trade that they had neglected during the messianic outburst. While the English merchants in Smyrna accused Sabbatai of working with Nathan of Gaza for 'neare 20 yeares' to cause 'great disturbances amongst the Jewes pretending to bee their Messiah',[119] they added that there 'is now greate hopes trade will suddenly much amend the Jews returning very eagerly again to their calling now willingly confessing their grosse errour'.[120] Because trade was the livelihood of the merchants, they soon

---

[114] *Oprechte Haerlemse Saterdaegse Courant*, 13 November 1666, as quoted in van Wijk, 'Wachtend op. de Wolk naar Jeruzalem', 66.

[115] *Oprechte Haerlemse Saterdaegse Courant*, 13 November 1666, as quoted in van Wijk, 'Wachtend op. de Wolk naar Jeruzalem', 66. The original stated: 'Het Schip de St. Victor, van Smirne komende, is alhier gearriveert, ende met de selve Brieven van daer van den 13 October, confirmerende, de genaemde Ioodse Messias Turckx geworden was'.

[116] *Oprechte Haerlemse Saterdaegse Courant*, 27 November 1666, as quoted in van Wijk, 'Wachtend op. de Wolk naar Jeruzalem', 67.

[117] Elisheva Carlebach, 'Sabbatianism and the Jewish–Christian Polemic', *Proceedings of the Tenth World Congress of Jewish Studies, Division C, Volume II: Jewish Thought and Literature* (Jerusalem: World Union of Jewish Studies, 1990), 1.

[118] Carlebach, 'Sabbatianism and the Jewish–Christian Polemic', 2.

[119] TNA SP 97/18/212: S. Pentlow, J. Foley and T. Laxton to T. Dethick, 29 September 1666.

[120] TNA SP 97/18/212: S. Pentlow, J. Foley and T. Laxton to T. Dethick, 29 September 1666. Meanwhile, the other merchants in Smyrna also spoke of their hopes that 'trade will revive for tis not to be expressed how farr they were gone in delusion for amongst themselves shown out all degrees of officers to sit in Jerusalem at their Restauration'. See TNA SP 97/18/210: A. Barnardiston, J. Adderley and N. Thurston to T. Dethick, 25 September 1666.

wrote with excitement that the Jews 'now begin to selle and promise to follow trading as before which they had totally neglected'.[121] Even in the Low Countries, a Dutch gazette commented on the economic impact of Sabbatai's apostasy: 'sijn Religie verlochent, en die van de Turcken aengenomen, om 't leven te behouden, waer mede nu de Joden weder tot stilte ghebracht zijn, en beginnen te Negotieren als te vooren' (he betrayed his religion, and adopted that of the Turk to preserve his life, wherefore also now the Jews are brought again to silence and begin to trade as before).[122]

The most popular account of the Sabbatian movement,[123] written by the English consul in Smyrna Paul Rycaut after the apostasy, presented the messiah and his supporters negatively and focused on the commercial effects as well. Rycaut referred to Sabbatai as 'their False messiah' and his followers as 'this Deceived People'.[124] Sabbatianism was 'a strange height of madness amongst the Jews' in which they fasted until they died, gave themselves lashes and buried themselves naked in their gardens.[125] For Rycaut, it was problematic that the Sabbatian believers buried themselves naked in their gardens because of the mercantile consequences of their actions: he pointed out that, when the Jews were performing ritual mortification, their shops were closed. Even in reference to the Jewish restoration to the holy land that many Christian millenarians longed for, Rycaut wrote: 'I perciev'd a strange transport in the Jewes, none of them attending to any business unless to winde up former negotiations, and to prepare themselves and Families for a Journey to Jerusalem'.[126]

Christian writers also saw the failure of the Sabbatian movement as a prime moment to push for conversion. Now that Sabbatai was proven a deceiver, they expected the Jews to realise the folly in their ways and embrace the true faith. The conversionist agenda was most pronounced in the *Lettera Mandata da Costantinopoli a Roma Intorno al Nuovo Messia de gli Ebrei* (1667). Unlike Rycaut, who wrote that 'God permitted the devil to delude this people',[127] this author did not ascribe diabolical origins to the Jewish movement. The Jews were just easily fooled by Sabbatai's cheap tricks. When Sabbatai was having his clothes changed to be dressed more appropriately as a 'Turk', they found three pounds of biscuits in his trousers that the Jewish messiah

---

[121]   TNA SP 97/18/211: A. Barnardiston, J. Adderley and N. Thurston to T. Dethick, 9 October 1666.

[122]   *Oprechte Haerlemse Saterdaegse Courant*, 13 November 1666, as quoted in van Wijk, 'Wachtend op. de Wolk naar Jeruzalem', 66.

[123]   Anderson, *An English Consul in Turkey*, 44, 215. It would be printed in at least 27 editions in English, French, German, Dutch, Polish, Russian and Welsh. For more on Paul Rycaut, see C.J. Heywood, 'Sir Paul Rycaut, A Seventeenth-Century Observer of the Ottoman State: Notes for a Study', in Ezel Kural Shaw and C.J. Heywood, eds, *English and Continental Views of the Ottoman Empire, 1500–1800* (Los Angeles: University of California, 1972), 33–59.

[124]   Evelyn, *History of the Three Late Famous Impostors*, 48, 52, 99.

[125]   Evelyn, *History of the Three Late Famous Impostors*, 78.

[126]   Evelyn, *History of the Three Late Famous Impostors*, 78.

[127]   Evelyn, *History of the Three Late Famous Impostors*, 62–3.

used to convince his followers that he could go for days without eating.[128] From the apostasy, the author turned to his hoped-for outcome of the messianic movement:

> Without doubt, that the stubbornness of the Jews was not in part an effect of the maledictions which God had given to that race, they would find a very reasonable motive to convert to Christianity perhaps which those who lived among you in Rome, and who have more comfort in easy education, than those of this country will obtain of the confusion that take of receiving in this case a powerful argument to open their eyes to truth; to come out of this deceits once after they have been deceived many times with vain hope, in which they persist to see being born for them some other liberator and messiah than the one whom we know in the venerable person of Jesus Christ.[129]

Sabbatai's apostasy afforded Christians with another opportunity not directed at the Jews. The failure of the Jewish messianic movement was used as a tool by some Protestants to attack their millenarian brethren. Thomas Coenen, the Dutch Reformed chaplain in Smyrna, wrote a text about Sabbatai and his following after witnessing the messianic outburst firsthand. Coenen sent his 140-page manuscript entitled *Ydele Verwachtinge der Joden Getoont in den Persoon van Sabethai Zevi* (1669) back to Amsterdam to be printed, dedicating it to its financers, 'the Honoured Club of the Lords Directors of the Levantine Commerce and Ship Company in the Mediterranean with its Seat in Amsterdam'.[130]

While Coenen attacked Sabbatai as the last seducer in a long line of Jewish pseudo-messiahs who had only brought misfortune to the stiff-necked Jews who refuse to believe in Jesus,[131] he utilised the Jewish movement to indirectly criticise Christian enthusiasts. Eschatological excitement, whether Jewish or Christian, would have the same result: failure and disappointment. He pointed to their similarities, citing verses (such as Joel 2:2–9) that were used in the prophecies of both Jewish Sabbatians and

---

[128] TNA SP 120/116: *Lettera Mandata da Costantinopoli a Roma Intorno al Nuovo Messia de gli Ebrei* (Rome, 1667).

[129] TNA SP 120/116: *Lettera Mandata da Costantinopoli a Roma Intorno al Nuovo Messia de gli Ebrei* (Rome, 1667). The original stated: 'Non è dubbio, che se l'ostinatione de gl'Ebrei, non era in parte un'effetto della maledittione, che Dio hà dato à quella razza, trovarebbero quì un motivo molto ragionevole per pigliare il partito del Christianesimo forse, che quei che vivono trà di voi in Roma, e che hanno più commodità di essere instruiti, che quegli di questi paesi, caveranno della confusione, che tengono di ricevere in questo caso un'argomento potentissimo, per aprire gli occhi alla verità; e disingannarsi una volta dopo esser stati ingannati tante volte con la vana speranza, nella quale persistono di veder nascere per loro qualch altro Liberatore, e Messia, che quello, che noi riconosciamo nella persona adorabile di Giesu Christo. Con che & c'.

[130] Gerbern Oegema, 'Thomas Coenen's "Ydele Verwachtinge Der Joden", (Amsterdam, 1669) as an Important Source for the History of Sabbatai Sevi', in Peter Schafer, Margarete Schluter and Giuseppe Veltri, eds, *Jewish Studies between the Disciplines: Papers in Honor of Peter Schafer on the Occasion of his Sixtieth Birthday* (Leiden: E.J. Brill, 2003), 333. Coenen's letters survive and two of them addressed to his employer at the 'Levantsen Handel y de Navigatie in de Middlezee tot Amsterdam' refer to Sabbatai. See NA R.A. Lev. H. 39 Port. 1.03.01.

[131] Oegema, 'Thomas Coenen's "Ydele Verwachtinge Der Joden"', 334.

Christian Fifth Monarchists.[132] Moreover, he compared the Sabbatian prophetic outbursts to those of the Quakers. Although he only mentioned English movements because he did not want to engage Dutch millenarians in explicit debate, his readers would have made the connection to the Protestant groups closer to home.[133]

Christian writers also subtly attacked millenarians by blaming the Jewish movement on Christian prophecies. The English merchants in Smyrna wrote that the Sabbatian messianic outbreak occurred because 'the Xtians did foresee such strange revolutions which would happen in this yeare 1666 as likewise so that those Jewes in Xtian lande as with as other pls [places] were at one and the same time equally possessd with beliefe'.[134] According to them, 'the Jews themselves say that nothing made them so willing to believe as the Friar predictions on the yeere 1666'.[135]

Even Paul Rycaut stated that the Jews, 'this subtle people judged this Year [1666] the time to stir, and to fit their Motion according to the season of the Modern Prophecies'.[136] Rycaut did not blame the false messiah or his Jewish supporters. It was the fanatical Protestant enthusiasts who dreamed of the restoration of the Jews, the inauguration of the fifth monarchy and the downfall of the pope in the year 1666 that were at the heart of the Jewish 'delusion'.[137] For this reason, it should not be surprising that Rycaut's pamphlet about the Sabbatian movement was reprinted multiple times in states that witnessed a rise of Christian millenarianism. It served as a reminder and a warning of the dangers of millenarian and messianic excitement.[138]

In sum, the apostasy of Sabbatai was seen by Christians in a fundamentally different light than Jews. For them, it was the end of a tragedy that offered hope that the Jews would return to their economic livelihoods and convert to Christianity. At the same time, it gave Christians a tool in their battle against millenarianism.

The Twist after the Twist

Although the disarray of Sabbatai's followers was indescribable, the messianic excitement had been so great that the apostasy could not quash it entirely. Sabbatai himself vacillated in his behaviour after the conversion. On one hand, he continued to write letters to Jewish communities, signing his name, 'Messiah of the God of Israel

---

[132]   Rapoport-Albert, *Women and the Messianic Heresy of Sabbatai Zevi*, 16n.

[133]   Heyd, 'The "Jewish Quaker": Christian Perceptions of Sabbatai Zevi as an Enthusiast', 242.

[134]   TNA SP 97/18/212: S. Pentlow, J. Foley and T. Laxton to T. Dethick, 29 September 1666.

[135]   TNA SP 97/18/211: S. Barnardiston, J. Adderley and N. Thurston to T. Dethick, 9 October 1666. While they had been led astray by Christian beliefs, the Jews of Smyrna were apparently indebted to the Europeans because 'they [the Jews] of this citye doe now confess to be beholding to the franks conversation, whose dissuading arguments have kept them from proving as mad as those have been at Salonica and other pls [places]'. See TNA SP 97/18/212: S. Pentlow, J. Foley and T. Laxton to T. Dethick, 29 September 1666.

[136]   Anderson, *An English Consul*, 24. This quote appears in numerous sources, including Evelyn, *History of the Three Late Famous Impostors*.

[137]   McKeon, 'Sabbatai Sevi in England', 135.

[138]   For more on this argument, see Popkin, 'Christian Interest and Concerns about Sabbatai Zevi'.

and Judah, Sabbatai Sevi'.[139] On the other, he would sometimes act as a pious Muslim and revile Judaism.

In the years after his conversion, Sabbatai became close to the Bektashi Sufi leader Muhammad Niyazi, who had developed a belief system that merged Bektashi, Jewish, Christian, Gnostic and Shiite concepts. Sabbatai most likely met the Sufi leader in 1666, when they attended some of the same dervish monasteries and prayer cells, or in 1669, when Niyazi stayed for 40 days at the court in Adrianople where Sabbatai was officially posted to meet Muslim dignitaries.[140] As Sabbatai made friends with Muslims, some of his Jewish followers embraced Islam at either Sabbatai's insistence or of their own volition. These people founded a small sect of Muslims, called the Donmeh, who secretly await the return of Sabbatai to this very day.

Among the Jews, Sabbatianism went underground. Some of Sabbatai's followers persisted in their Jewish faith and their belief in the messiah. In particular, post-apostasy Sabbatianism found a particularly strong response among former Conversos who had been among the most committed followers and saw a repetition of their own families' histories in the messiah's action. The Conversos' dialectic with Christianity meant that they were preoccupied with the identity of the messiah but not deeply rooted in the messianic tradition of either Christianity or Judaism. This created an unusual flexibility in the variety of messianic scenarios that they were willing to entertain. Some of these featured the belief that the messiah would be a Converso. They were therefore ready to accept the necessity of Sabbatai's conversion and came up with ingenious interpretations to convince themselves that the religious writings in the biblical, rabbinic and kabbalistic literature foretold his apostasy.[141] They came to the conclusion that the messiah must be a Converso to Islam in order to redeem the world.[142] The conversion was just an outward mask necessary to cover a different inner experience. In other words, Sabbatai was living a life similar to that of a crypto-Jewish Converso.[143]

This link was made explicit in the words of the Sabbatian prophet Abraham Cardoso, who exclaimed in a letter to his brother Isaac in 1668:

---

[139]   Ben Zvi Institute facsimile of MS 2262/79: Sabbatai Sevi to the Jewish community of Berat, August 1676.

[140]   Fenton, 'Shabbetay Sebi and His Muslim Contemporary Muhammed an-Niyazi', 82–3.

[141]   Sharot, *Messianism, Mysticism, and Magic*, 117, 119. While the life experience of former Conversos could work in favour of their continued belief in Sabbatai Sevi, as Schacter notes, there 'were also equally compelling factors which militated against such a belief'. For more, see Jacob Schacter, 'Motivations for Radical Anti-Sabbatianism: The Case of Hakham Zevi Ashkenazi', in Rachel Elior, ed., *Sabbatianism and Its Aftermath: Messianism, Sabbatianism, and Frankism, Vol. II* (Jerusalem: Institute of Jewish Studies, 2001), 31–50, 33 and Yerushalmi, *From Spanish Court to Italian Ghetto: Issac Cardoso*, 302–49.

[142]   Goldish, *The Sabbatean Prophets*, 58.

[143]   Goldish, *The Sabbatean Prophets*, 49, 46; Yerushalmi, *From Spanish Court to Italian Ghetto: Isaac Cardoso*, 304.

it was also two years ago that it was told me that the king messiah was destined to wear the clothes of a *converso* [anus], because of which the Jews would not recognize him; and in fine, that he was destined to be a *converso* like me.[144]

By connecting the messiah's conversion to the Converso experience, Cardoso justified Sabbatai's actions and gave the former Conversos' acceptance of Christianity a religious and messianic significance.[145]

Abraham Cardoso, in particular, was an ideal propagandist for the apostate Sabbatai because of his own experience as a Converso. Cardoso had studied Christian theology in Spain before he became a rabbi. His earlier beliefs had a lasting impact. He understood Judaism in light of his Converso heritage, never fully divorced his new Jewish beliefs in Sabbatai from his Christian background and developed a theology that synthesised Jewish and Christian messianic ideas. He even applied the same biblical verses to Sabbatai that Christians cited for Jesus,[146] believing that the suffering servant of Isaiah referred to the messiah (as Christians claimed) at the same time that he harnessed the two-messiah theory to state that it referred to the messiah of the house of Joseph.[147]

In Amsterdam, the Christian chiliast Petrus Serrarius supported the few Jews who remained faithful to Sabbatai. Serrarius was Sabbatai's 'first and leading' Christian supporter both before and after the apostasy.[148] While one may think that 'a believing Christian could hardly be a fully believing Sabbatian,'[149] Serrarius accepted both Sabbatai's messiahship (with the two-messiah theory) and the rationalisation of the apostasy. God works in mysterious ways, Serrarius quipped.

William Sherwin was another Christian who saw the Sabbatian movement, even after the messiah's apostasy, in a millenarian light. He referred to Sabbatai's failed messiahship as a premonition of Christ's return: 'the neer approach of whose said coming ... the world of late years hath been sufficiently alarm'd, by the Jews attempt ... to seek their promised Land'.[150]

---

[144]   Abraham Cardoso to Isaac Cardoso in 1668, as translated and quoted in Goldish, *The Sabbatean Prophets*, 99.

[145]   Yosef Kaplan, *From Christianity to Judaism: The Story of Isaac Orobio de Castro* (Oxford: Oxford University Press, 1989), 213.

[146]   Yerushalmi, *From Spanish Court to Italian Ghetto: Isaac Cardoso*, 338–9; Scholem, *Major Trends in Jewish Mysticism*, 309–10.

[147]   Goldish, *The Sabbatean Prophets*, 44, 49. For more on Cardoso's messianism, see Bruce Rosenstock, 'Abraham Miguel Cardoso's Messianism: A Reappraisal', *Association of Jewish Studies Review* Vol. 23, No. 1 (1998), 63–104. For more on Cardoso in general, see Yerushalmi, *From Spanish Court to Italian Ghetto: Isaac Cardoso*.

[148]   Van der Wall, 'The Amsterdam Millenarian Petrus Serrarius (1600–1669) and the Anglo-Dutch Circle of Philo-Judaists', 90.

[149]   Goldish, *The Sabbatean Prophets*, 155.

[150]   William Sherwin, *The True News of the Good New World Shortly to Come* as quoted in Warren Johnston, *Revelation Restored: The Apocalypse in Later Seventeenth-Century England* (Woodbridge: The Boydell Press, 2011), 10.

Returning to the Dutch Republic, the Protestant scholar Serrarius proved his dedication when he set out from the Dutch Republic towards the Ottoman Empire two years later in hopes of meeting the Jewish messianic convert to Islam. Although Serrarius died en route to Adrianople, never meeting the man who had captivated his attention for the last few years, his beliefs came to include both Sabbatai and the universal redemption of Jews, Christians and Muslims. Indeed, Serrarius was one of the proponents of Jewish–Christian–Islamic unity that became increasingly common among 'Menasseh's most famous interlocutors'.[151]

Studying the narratives about Sabbatai written by Christians adds a vital dimension to our knowledge of the cross-religious representations of the Sabbatian movement. In particular, this chapter has followed Michael Heyd who, later in his career, turned his attention to the 'underexamined attitudes of Christian European thinkers toward non-Christian phenomena of religious enthusiasm, such as the messianic movement of Sabbatai Zevi'.[152] While Heyd and other historians, such as Gershom Scholem, Michael McKeon, Jetteke van Wijk, Ernestine van der Wall, Richard Popkin, Ingrid Maier and Daniel Waugh, have examined Christian interest in the Jewish movement, none of them apparently knew about the Italian avvisi or the correspondence of English merchants and diplomats in Tuscany and the Low Countries that contained references to Sabbatai Sevi.

Despite the variety and fluidity of Christian attitudes towards the Jewish messiah, academics have homogenised this diversity and tended towards extremes. On the one hand, some have argued that Jewish messianists and Christian millenarians shared in their hopes towards Sabbatai for entirely different reasons – even if the Christian views inevitably reinforced the Jewish beliefs.[153] On the other hand, some have made claims about cross-religious influences that are unfounded and have gone too far by declaring that Comenius, Oldenburg, Labadie and Serrarius were all complete Sabbatian believers or 'Jewish Christians'.[154] In reality, these men held differing perspectives that changed over time, and we do not know enough about Comenius' beliefs to say such things.[155]

Oldenburg, for example, can serve as an example of evolving Christian beliefs in Sabbatai because there appears to be three phases in his reaction to the Sabbatian movement. At the beginning, he attached great hopes to it. After the messiah's conversion, Oldenburg lost faith in the messianic revolution even though he continued to gather news. Finally, he came to a point of 'apocalyptic apprehension'

---

[151] Katchen, *Christian Hebraists and Dutch Rabbis*, 104.

[152] Tamar Herzig, 'Introduction', in Asaph Ben-Tov, Yaacov Deutsch and Tamar Herzig, eds, *Knowledge and Religion in Early Modern Europe: Studies in Honor of Michael Heyd* (Leiden: Brill, 2013), 2.

[153] Lionel Kochan, *The Making of Western Jewry, 1600–1819* (New York: Palgrave Macmillan, 2005), 144.

[154] Richard Popkin, 'Christian Jews and Jewish Christians in the 17th Century', in Richard Popkin and G.M. Weiner, eds, *Jewish Christians and Christian Jews: From the Renaissance to the Enlightenment* (Dordecht: Kluwer Academic Publishers, 1994), 68–9.

[155] Van der Wall, *De Mystieke Chiliast Petrus Serrarius (1600–1669) en Zijn Wereld*, 407.

where he noted that few people believed in the imminent messianic age but many still hoped it would come soon.[156]

This chapter has sought to present the wealth of Catholic and Protestant reactions in a more nuanced manner and, specifically, in the way that they evolved. Upon first hearing of the Jewish messianic outburst, Christians responded in multiple ways. Many dismissed it outright. Others were more objective. Meanwhile, the growth of philo-Semitism and Judeocentric millenarianism meant that some Protestants hoped that the Jewish restoration would actually occur; they co-opted this element of the Jews' beliefs and placed it within their Christian eschatological schema. A select few, such as Serrarius, went further. They accepted Sabbatai's messiahship, utilising the two-messiah theory to mesh the advancing Jewish messianic spirit with their Christian millenarian one.

With the messiah's conversion to Islam at the height of his popularity, most of the Jews abandoned their faith in Sabbatai as quickly as they had embraced him. At the same time, Christians predominately looked down on the Jewish movement after the apostasy, hoping that its failure would lead the Jews to convert to Christianity and return to their business and trade. But for Serrarius and others like him, Sabbatai's mission was not yet finished. They accepted the rationalisation of his conversion to Islam, incorporating Jews, Christians and now Muslims into their eschatological expectations.

---

[156]   For more, see Loker, 'English Contemporary Opinions on the Sabbatean Movement'.

Conclusion

Coming full circle to conclude where we began, Peter Serrarius was setting out on a journey from the Dutch Republic to the Ottoman Empire in hopes of meeting Sabbatai Sevi. To understand why this Protestant scholar was travelling across an entire continent to meet a Jewish convert to Islam that he had neither met nor corresponded with before, one must discuss Serrarius within his broader intellectual and geographical environment.

Serrarius' beliefs were part of the growth in eschatological excitement between 1648 and 1666 in which apocalyptic ideas originating in the American colonies in the 1640s were transmitted throughout the European states to the Ottoman Empire in the 1650s, preparing the minds of Jews and Christians for the return of such narratives from the Middle East in the 1660s. This three-stage process had an intellectual legacy that stretched back to 1492. The traumatic experience of exile, persecution and relocation associated with the expulsion of the Jews from Spain, the forced conversions in Portugal, the splintering of Christendom with the Reformation and the European discovery of the Americas were interpreted as the birth pangs of the messianic age and signalled the coming end. The very same historical events and processes that encouraged this apocalyptic tension spawned migration that created an infrastructure of networks, which facilitated the transnational and cross-religious transmission that were vital in the movement of eschatological constructs in the seventeenth century.

The expansion of apocalyptic beliefs during this period began in the 1640s when the Converso Antonio de Montezinos arrived in Amsterdam with an incredible tale of the discovery of the Lost Tribes in the jungles of South America. Menasseh ben Israel shared Montezinos' narrative with the Sephardic community to which he belonged and whom he taught as well as his philo-Semitic friends in England. In Europe, the Conversos' testimony was used as a tool to promote Jewish messianism, Christian millenarianism and the readmission of the Jews to England. In the Mediterranean world, Menasseh's *Hope of Israel* was republished in Smyrna by Sabbatai's friends in the years of his youth, and it possibly informed Sabbatai's messianic yearnings. In the Atlantic world, it was a stimulus for the growth of apocalyptic interpretations of the American aboriginals' history, which affected the beliefs of John Eliot, the Puritan missionary in New England. Indeed, to understand how Eliot's views of the Lost Tribes changed multiple times between 1648 and 1666, one must consider the stories and theories about the Israelites that were spread back and forth across the seventeenth-century Abrahamic world.

The growing eschatological excitement in England was felt in Quaker circles and encouraged James Nayler's messianic entrance into Bristol in 1656. While Protestants and Catholics across Europe knew of Nayler's actions, the lack of a news industry in the Ottoman lands meant that news of European events transmitted to the Empire by merchants, diplomats and other travellers was only disseminated orally among smaller populations. There were no avvisi, gazettes, pamphlets or newsbooks to spread this information to a larger audience: an asymmetry between the developments of news media at opposite ends of this chain of transmission.

The case of Nayler also points to the asymmetrical relationship between Judaism and Christianity. The messianism associated with James Nayler neither reached Sabbatai or Nathan nor served as the impetus for the Sabbatian movement. Indeed, this disjuncture highlights the long-standing trend in which incidents in Christendom did not affect the Jews to the same extent that Jewish movements influenced Christians. After all, Judaism was part of the Christian heritage, but not vice versa.

Because the Lost Tribes had been incorporated into the Christian eschatological sequence, many millenarians expected the reappearance of the ancient Hebrews. When the rumours about the re-emergence of the Lost Tribes in 1665 reached Europe, they therefore stimulated Christian millenarian hopes and reinforced the belief that the second coming of Jesus was at hand. At the same time, these tales inspired Jewish messianic excitement and set the stage for Sabbatianism.

The widespread transmission of the narratives about the Israelites demonstrates the intertwining of Jewish, Christian and Islamic eschatological beliefs against the background of an increasingly global exchange of news and rumours. The emerging global communication infrastructure facilitated the circulation of reports, but it did not provide a ready means for people to ascertain their veracity. This meant that rumours, including those of an apocalyptic nature, reached new heights and had a greater impact than ever before. Interest in the Lost Tribes would probably never have reached the intensity that it did were it not for the vast distances separating Europe from both South America and the Arabian Peninsula, which allowed the spread of claims that could rarely be either substantiated or disproved.

The religious constructs that transcended these spaces, connecting people and shrinking their perceptions of the world, were more readily accepted because of the long way that they travelled. The early modern world was so full of new claims – about new continents, people unmentioned in the scriptural record, whole civilisations unknown to classical philosophers and astounding new kinds of plants, animals, artefacts and technologies – that it seemed impossible to determine what was possible, what was unlikely and what was true. These distances also forced people who wanted to accept the truth of what they heard to do so relying primarily on hearsay. Even Eliot, who lived in New England, had to rely on the supposed claims of a dead man in North America for corroborating evidence for the Lost Tribes theory.

The size of the Atlantic and Mediterranean was palpable, especially for people in Europe. Thorowgood's eagerness to publish Eliot's firsthand report demonstrates that he saw the gap between the two places as too large to cross and verify the information for himself. Although the space of the Atlantic was perceived as vast from a European vantage point, the same men did not perceive distances on the other side of the Atlantic as large at all. Eliot was seen as a source on aboriginals who lived far away from him, and Menasseh and Thorowgood cited examples of aboriginal practices throughout the Americas, conflating both the peoples and the distances that separated them. In the Mediterranean world, Mecca was one of the hardest cities for Europeans to visit and they knew so little about it that many incorrectly thought Muhammad was buried there. More surprising, there was a disjuncture between Egypt and the Hijaz: Jews in Alexandria were not aware of the political structure in the Arabian Peninsula, which was in their own state.

Even the geography of the early modern world was conducive to the growth of messianism and millenarianism. The expansion of exploration and migration created the right combination of hidden space to promote eschatological excitement. Unlike the medieval and modern periods, the early modern period had enough unknown areas where the Lost Tribes could be hiding as well as enough people going towards these locations and enough communication coming back from them to allow for the proliferation of rumours about the Israelites that could not readily be disproved. In the medieval period, no European knew of South America. In the modern age, satellites and global positioning systems can pinpoint the places where people claimed the ancient Hebrews were dwelling. Therefore, the very space of the world in this period coupled with the manner in which people were crossing it created an environment in which apocalyptic excitement thrived.

Finally, Serrarius' attempted pilgrimage to Adrianople in 1669 was the direct result of the rise of the Jewish Sabbatian movement in 1666 that captivated the attention of Christians across the Mediterranean and Atlantic worlds. Catholics and Protestants of a variety of occupational, religious and national backgrounds sought to comprehend the Jews' beliefs in their new messiah when his followers began abandoning their businesses and planning to return to the holy land. The letters, newsletters, pamphlets, gazettes, newsbooks and full-length publications about Sabbatai written in Italian, Dutch, Latin, English, French and German informed populations across Christendom of the latest news of the Jewish messianic movement in the Levant.

Indeed, the advent of the news industry and news culture enabled the spread of the stories about the messiah in many forms to broader populations, encouraging the rise of messianic and millenarian excitement. The emergence of the widespread dissemination of news from abroad was a development in this period that affected the way that many people understood the world around them. For some, news became an obsession. As Roger Williams noted in the seventeenth century, 'The whole race of mankind is generally infected with an itching desire of hearing Newes'.[1]

Apocalyptic constructs were also transmitted along scientific networks that made their appearance in the seventeenth century with the inauguration of institutions such as the Royal Society. The founders of the Royal Society were serious scholars, working on scientific projects, who were also fascinated with the stories about the end of the world. The Society was an arena in which academic and eschatological ideas were disseminated. We may see information in letters from friends, in newspapers,

---

[1] Williams, *A Key to Language of America*, 61.

in academic dialogue and in religious texts as belonging to different categories of knowledge, but these individuals believed that the scriptures outlined the entire course of history; events were expected to fit into the timeline presented in the biblical books. Like many others, they read the news that they received in the gazettes, avvisi, pamphlets and correspondence in relation to their sacred texts. In their minds, science, news and religious beliefs came together in a manner that promoted eschatological excitement.

While many Christians were against Sabbatai and his followers from the start, the growth of philo-Semitism and Judeocentric millenarianism meant that some Protestants were ready to accept and co-opt the Jews' messianic hopes, reframing them in terms of their own Christian views. A few of these people, such as Serrarius, used the two-messiah theory to fuse elements of Jewish messianism together with Christian millenarianism.

After Sabbatai apostatised to Islam at the height of his popularity, most Christians and Jews considered him as nothing more than an impostor. Serrarius, on the other hand, remained Sabbatai's leading Christian supporter, accepting the rationalisation of the messiah's conversion. Although Serrarius died en route to Adrianople in 1669, the momentary Jewish messianic outburst had a substantial impact on the development of the Dutch chiliast's religious expectations. By the time that he died, Serrarius believed in universal redemption that included Jews, Christians and Muslims.

Historians know a good deal about the global flow of trade, but they are just beginning to consider the global circulation of information.[2] This text has sought to add to the growing body of literature in this field by tracing the three-stage process of eschatological transmission between 1648 and 1666 that highlights the interconnections in apocalyptic thinking across the seventeenth-century Abrahamic world. After all, one cannot understand Serrarius' 1669 journey or the increase of eschatological excitement in 1666 without examining the previous two decades of transnational transmission and cross-religious exchanges that connected Jews, Christians and Muslims from the Americas through Europe to the Levant.

---

[2]   Benton, 'The British Atlantic in Global Context', 272.

Appendix
Parallel Timelines (1614–1669)

Year	The Americas	Iberia and Italy	Northern Europe	The Levant
1614–47	1641: Antonio de Montezinos discovers the Lost Tribes in South America 1646: John Eliot begins his missionary work among the aboriginals in New England		1644: Montezinos shares his story with Menasseh ben Israel in Amsterdam c. 1645: John Dury hears about Montezinos' testimony from Menasseh	1614: Sabbatai Sevi's father moves from southern Greece to Smyrna, where he acquires a job as a factor for English merchants
1648: The year that kabbalists anticipated the Jews' redemption			Thomas Thorowgood gives a copy of his manuscript *Iewes in America* to Dury	Sabbatai's erratic behaviour in Smyrna culminates with his claim that he is the messiah
1649–55	1650: Eliot accepts the Lost Tribes theory after reading the manuscripts of Thorowgood, Menasseh and others sent to him by Winslow The growth of millenarianism leads some New Englanders to return to England to join the Fifth Monarchy Men 1655: The beginning of Quaker missions to New England via Barbados		1649: Edward Winslow publishes *The Glorious Progress of the Gospel* with the help of Dury 1650: Menasseh publishes the *Hope of Israel* 1650: Thorowgood publishes *Iewes in America* 1654: Menasseh introduces the two-messiah theory to English Protestants 1655: The campaign for the readmission of the Jews to England	1651–54: Sabbatai is expelled from Smyrna for his blasphemous behaviour and wanders throughout the Ottoman Empire

Year	The Americas	Iberia and Italy	Northern Europe	The Levant
1656: The year that English Protestants expected the conversion of Jews, and the return of Jesus and the onset of the millennium	October: Eliot withdraws his support for the Lost Tribes theory when the anticipated millennium does not begin		October: James Nayler's messianic entrance into Bristol	
1657–64		1657: Nayler is discussed in Italian avvisi 1659: Maurice Conry writes about Nayler in *De Extremis Anglo-Haereticorum* after being imprisoned with him in England	1657: The Dutch *Hollandtze Mercurius* and *Klachte der Quakers* report on Nayler's actions and punishment 1659: Nayler is released from prison 1660: Nayler dies in Great Britain 1660: Thorowgood republishes *Jews in America* with a statement from Eliot 1662: The founding of the Royal Society	1657: The first Quaker mission to the Ottoman Empire via Livorno 1657: A Quaker pilgrim arrives in Jerusalem while Nathan of Gaza is there *c.* 1658: Abraham Gabbai reprints the *Hope of Israel* in Smyrna 1658: The Quaker missionary Mary Fisher and Sabbatai may have been in Istanbul at the same time 1661: Quaker missionaries and Sabbatai are both in Smyrna 1662: Quaker missionaries in Alexandria may have crossed paths with Sabbatai

Year	The Americas	Iberia and Italy	Northern Europe	The Levant
1665		April: The rumoured sack of Mecca by solely Arabs first appears in a Venetian avviso		April: Nathan has a vision that Sabbatai is the true messiah
	June: New Amsterdam becomes New York			June: Nathan anoints Sabbatai as the messiah
			July: Dutch pamphlets print stories of the Lost Tribes sacking Mecca	
		August: A Venetian avviso reports that the Arabs at Mecca are joined by a *numero infinito* of Jews		Summer: Unknown English merchant in Smyrna meets Sabbatai and writes to his English associates in Tuscany about the visit
	September: John Davenport is aware of the Jewish movement in the Levant		July: The supposed destruction of Mecca is known about in Royal Society circles	
	Winter: Increase Mather preaches a series of sermons in Boston on Revelation that are inspired by the news that the Jews are heading to Palestine		December: The English population reads about the Lost Tribes sacking Mecca and Sabbatai's messiahship in the *Gazette*	
			December: The Sabbatian movement becomes a popular topic in Dutch gazettes due to its economic effects	

Year	The Americas	Iberia and Italy	Northern Europe	The Levant
1666: The year that Jews and Christians expected the imminent end of history and their ultimate redemption	February: Increase Mather delves into the scriptural writings and the political history of the Jews in the Levant Summer: Mather tries to convince three Jews that the messiah has already come	April: Englishmen in Tuscany write reports about Sabbatai June: Stories about the Jews' saviour appear in the Italian press December: The story of Sabbatai's conversion is printed in a Turin avviso	November: News about the apostasy reaches English and Dutch presses	February: Sabbatai is imprisoned in Istanbul March: The Venetian bailo mentions the Jewish messiah in his diplomatic dispatch September: Sabbatai converts to Islam in Adrianople October: English merchants in Smyrna write about Sabbatai's conversion in letters to their associates in Tuscany
1667–69		1667: An Italian broadsheet is printed in Rome that provides the entire history of the Sabbatian movement	1667: Some Jews remain in denial about the messiah's conversion 1669: John Evelyn's *A History of the Three Late Famous Impostors* is published in London 1669: Petrus Serrarius sets out on his journey from Amsterdam to meet Sabbatai in Adrianople	1667: Thomas Coenen completes his Dutch account of the outbreak of Sabbatianism 1669: Sabbatai becomes friends with the Sufi leader Muhammad Niyazi in Adrianople

Select Bibliography

Archival Sources

Archivio di Stato di Venezia (Venice State Archives), Venice
Senato Dispacci Ambasciatori Constantinopoli F. 150: Giovan Battista Ballarin to the Venetian doge and senate.

Archivio Segreto Vaticano (Vatican Secret Archives), Vatican City
Segreteria di Stato, Avvisi: avvisi from Genoa, Turin and Venice.

American Antiquarian Society, Worcester
Fletcher, Giles. *Israel Redux; or the Restauration of Israel* (London, 1677).
Jews Jubilee; or the Conjunction and Resurrection of the Dry Bones of the House of Israel (London, 1688).
Mather's Family Papers Box 3, Folder 1: Increase Mather's diary.

Beinecke Rare Books and Manuscripts Library, University of Yale, New Haven
Carlingford Papers Box 2: William Temple's correspondence.

Ben Zvi Institute, Jerusalem
Facsimile of MS 2262/79: Sabbatai Sevi to the Jewish community of Berat, August 1676.

Biblioteca Nazionale Centrale di Firenze (Central National Library of Florence), Florence
Codd Magliabechiani XXV: avvisi from Genoa, Venice and Milan.

Bibliotheca Rosenthaliana, University of Amsterdam, Amsterdam
Anabaptisticum et Enthusiasticum Pantheon und Geistliches Rust-Hauss (Hamburg, 1702).
Coenen, Thomas. *Ydele Verwachtinge der Joden Getoont in den Persoon van Sabethai Zevi* (Amsterdam, 1669).
Croesi, Gerardi. *Historia Quakeriana* (Amsterdam, 1695).
Klachte der Quakers, Over Haren Nieuwen Marterlaer, James Nailor in Engelandt (1657).

The Jewish Theological Seminary, New York
Serrarius, Peter. *The Congregating of the Dispersed Jews* (London, 1666).
_____. *Gods Love to His People Israel* (London, 1666).

London Society of Friends' Library, London
Box A/4: James Nayler's correspondence.
Caton MS: Richard Hubberthorne's letters to Margaret Fell.
Edmundson, William. *A Journal of the Life, Travels, Sufferings, and Labour of Love in the Work of the Ministry, of that Worthy Elder and Faithful Servant of Jesus Christ, William Edmundson* (Dublin, 1820).
Green, Joseph Joshua, *Biography of Samuel Cater of Littleport in the Isle of Ely, Yeoman*. Typewritten copy, 1914.
The Letter Sent by Robert Rych to William Bayly and Mary Fisher, called his Wife and To the rest of the Quakers Hearers and Followers (London, 1669).
Pamphlet Box L. 24: Lacock, Bettina. 'Quaker Missions to Europe and the Near East 1655–1665'. Undergraduate thesis, Birmingham University, 1950.
Perrot, John. *The Blessed Openings of a Day of Good Things to the Turks*. (1661).
Port MS: correspondence of John Luffe, John Perrot and John Stubbs.
Swarthmore MS I–V: letters from John Stubbs, George Bishop and John Perrot.

Massachusetts Historical Society, Boston
Richards-Child Family Papers MS: John Hull's diary.

Nationaal Archief (National Archives), The Hague
R.A. Lev. H. 39 Port. 1.03.01: Thomas Coenen's letters.

New York Public Library, New York
*KC 1666: Maraschalck, Lira. *A Brief Relation of Several Remarkable Passages of the Jevves* (London, 1666).

The National Archives, London
SP 29/162: letter from Thomas Chappell.
SP 97/18: letters from English merchants and diplomats in the Ottoman Empire.
SP 98/6–7: correspondence of the English colony in Tuscany.
SP 120/116: *Lettera Mandata da Costantinopoli a Roma Intorno al Nuovo Messia de gli Ebrei* (Rome, 1667).

University Library of Leiden, Leiden
Castelyn, Pieter, ed. *Hollandtze Mercurius, Behelzende de Gedenckweerdichste Voorvallen in 't Jaer 1665* (Haarlem, 1666).
_____, ed. *Hollandtze Mercurius, Vervatende het Gepasseerde in Europa: Voornamentlijck in den Engelze ende Nederlantschen Oorlog, in 't Jaer 1666* (Haarlem, 1667).
_____, ed. *Hollandtze Mercurius, Vervatende de Voornaemste Gelchiedenisse Voorgevallen in 't Christenrijck in 't Jaer 1656* (Haarlem, 1657).

Historis Verhael van den Nieuwen Gemeynden Koning der Joden; Sabatha Sebi, als Mede sijn by Hebbende Propheet Nathan Levi. (1665).

Other Primary Documents

Archer, John. *The Personal Reign of Christ upon Earth* (London, 1642).

Baker, Daniel. *A Clear Voice of the Truth Sounded Forth* (London, 1662).

Barnardiston, Giles. 'Abstract of Will of Giles Barnardiston', *Journal of the Historical Society of Friends* Vol. 7 (1910), 43–4.

Ben Israel, Menasseh. *The Hope of Israel* (London: Livewell Chapman, 1652).

_____. *Vindicae Judaeorum or a Letter in Answer to Certain Questions Propounded by a Noble and Learned Gentleman* (London, 1656).

Besse, Joseph, ed. *A Collection of the Sufferings of the People called Quakers for the Testimony of a Good Conscience*, 2 Volumes (London, 1753).

Bishop, George. *The Throne of Truth Exalted over the Povvers of Darkness* (London, 1657).

Brennan, Michael. *The Origins of the Grand Tour: The Travels of Robert Montagu, Lord Manderville 1649–1654; William Hammond, 1655–1658; Banaster Maynard, 1660–1663* (London: The Hakluyt Society, 2004).

Bruodin, Anthony. *Propugnaculum Catholicae Veritatis* (Prague, 1669).

Calder, Isabel MacBeath, ed. *Letters of John Davenport: Puritan Divine* (New Haven: Yale University Press, 1937).

Calendar of State Papers Domestic Series ... preserved in ... Her Majesty's Public Record Office, 23 Volumes (London: Longman and Co., 1858–1897).

Cary, Mary. *The Little Horns Doom and Downfall* (London, 1651).

Cotton, John. *The Churches Resurrection* (London, 1642).

_____. *An Exposition upon the Thirteenth Chapter of Revelation* (London, 1656).

_____. *The Powring Out of the Seven Vials* (London, 1642).

Croese, Gerardus. *The General History of the Quakers* (London, 1696).

Crossley, James, ed. *The Diary and Correspondence of Dr. John Worthington* (London, 1847).

Deacon, John. *The Grand Impostor Examined* (London, 1656).

A Door of Hope: Or, A Call and Declaration for the Gathering Together of the First Ripe Fruits unto the Standard of our Lord King Jesus (London, 1660).

Dury, John. *An Information Concerning the Present State of the Jewish Nation in Europe and Judea* (London: R.W., 1658).

Evelyn, John. *History of the Three Late Famous Impostors* (London, 1669).

Farmer, Ralph. *Sathan Inthron'd in his Chair of Pestilence* (London, 1656).

Finch, Henry. *The Worlds Great Restauration* (London, 1621).

Freiberg, Malcolm, ed. *Winthrop Papers*, 6 Volumes (Boston: Massachusetts Historical Society, 1992).

Goodwin, Thomas. *A Sermon of the Fifth Monarchy* (London, 1654).

Hall, Rupert and Marie Boas Hall, eds. *The Correspondence of Henry Oldenburg*, 13 Volumes (Madison, Milwaukee and London: The University of Wisconsin Press, 1965–73).

Hinds, Allen, ed. *Calendar of State Papers Relating to English Affairs Existing in the Archives and Collections of Venice and in Other Libraries of Northern Italy* (London, 1930).

Kaplan, Yosef, trans. and ed. *Thomas Coenen's Vain Hopes of the Jews as Revealed in the Figure of Sabbetai Zevi* (Hebrew) (Jerusalem: The Ben-Zion Dinur Institute for Research in Jewish History, 1998).

La Peyrere, Isaac. *Men Before Adam* (London, 1656).

The Last Letters To the London-Merchants and Faithful Ministers (London, 1665).

The London Gazette (London, 1666–67).

Mather, Cotton. *The Life and Death of the Reverend Mr. John Eliot, who was the First Preacher of the Gospel to the Indians in America* (London, 1694).

_____. *Magnalia Christi Americana*, 2 Volumes (London, 1702).

Mather, Increase. *The Mystery of Israel's Salvation* (London, 1669).

Offenberg, Adri. 'Uit de Bibliotheca Rosenthaliana', *Studia Rosenthaliana* Vol. 29, No. 1 (1995), 91–9.

The Oxford Gazette (Oxford, 1665).

Publick Intelligencer (London, 1656–57).

Roth, Cecil, ed. *Anglo-Jewish Letters (1158–1917)* (London: The Soncino Press, 1938).

Serrarius, Peter. *An Awakening Warning to the Wofull World by a Voyce in Three Nations* (Amsterdam, 1662).

_____. *The Restauration of the Jews* (London, 1665).

Simonsohn, S. 'A Christian Report from Constantinople Regarding Shabbethai Sevi (1666)', *Journal of Jewish Studies* Vol. 12, No. 1–2 (1961), 33–58.

'Thomas Bendish's Report in the Calendar of State Papers', *Journal of the Friends Historical Society* Vol. 8 (1911), 168.

Thorowgood, Thomas. *Digitus Dei* (London, 1652).

_____. *Iewes in America, or, Probabilities that the Americans are of that Race* (London, 1650).

_____. *Jews in America* (London, 1660).

Vaughn, Thomas. *The Fame and Confession of the Fraternity of R: C: Commonly, of the Rosie Cross* (London, 1652).

Villani, Stefano, ed. *La Corrispondenza dei Residenti Toscani a Londra: Commonwealth e Protettorato (11 Dicembre 1648 – 11 Giugno 1660)*. Unpublished manuscript.

Whitfield, Henry. *The Light Appearing More and More Towards the Perfect Day* (London, 1651).

Whitlock, Bulstrode. *Annals of the Universe: Containing an Account of the Most Memorable Actions, Affairs, and Occurences which have Happened in the World but Especially in Europe, from the Year 1660* (London, 1709).

Williams, Oliver, ed. *Mercurius Politicus* (London, 1656–57).

Williams, Roger. *A Key to Language of America* (London, 1643).

Winslow, Edward. *The Glorious Progress of the Gospel amongst the Indians in New England* (London, 1649).

Secondary Studies

Abulafia, David. *The Discovery of Mankind: Atlantic Encounters in the Age of Columbus* (New Haven and London: Yale University Press, 2008).

_____. *The Great Sea: A Human History of the Mediterranean* (London: Penguin Books, 2011).

Akerman, Susanna. *The Rose Cross Over the Baltic: The Spread of Rosicrucianism in Northern Europe* (Leiden: E.J. Brill, 1998).

Aescoly, A.Z. 'David Reubeni in the Light of History', *The Jewish Quarterly Review* Vol. 28, No. 1 (1937), 1–45.

Alexander, Paul. *The Byzantine Apocalyptic Tradition* (Berkeley: University of California Press, 1985).

Altman, Ida and James Horn, eds. *'To Make America': European Emigration in the Early Modern Period* (Berkeley: University of California Press, 1991).

Anderson, M.S. *The Rise of Modern Diplomacy 1450–1919* (London: Longman, 1993).

Anderson, Sonia. *An English Consul in Turkey: Paul Rycaut at Smyrna, 1667–1678* (Oxford: Clarendon Press, 1989).

Arbel, Benjamin. *Trading Nations: Jews and Venetians in the Early Modern Eastern Mediterranean* (Leiden: E.J. Brill, 1995).

Arjomand, Said Amir. 'Islamic Apocalypticism in the Classic Period', *The Encyclopedia of Apocalypticism II* (New York: Continuum, 2000), 238–83.

Armitage, David and Michael Braddick, eds. 'Introduction', *The British Atlantic World, 1500–1800* (New York: Palgrave Macmillan, 2009), 1–12.

Baer, Marc. *The Donme Jewish Converts, Muslim Revolutionaries, and Secular Turks* (Stanford: Stanford University Press, 2009).

_____. 'Globalization, Cosmopolitanism, and the Donme in Ottoman Salonica and Turkish Istanbul', *Journal of World History* Vol. 18, No. 2 (2007), 141–70.

Bailyn, Bernard. *New England Merchants in the Seventeenth Century* (Cambridge, Mass.: Harvard University Press, 1955).

Barbarics, Zsuzsa and Renate Pieper. 'Handwritten Newsletters as a Means of Communication in Early Modern Europe', in Francisco Bethencourt and Florike Egmond, eds, *Cultural Exchange in Early Modern Europe III* (Cambridge: Cambridge University Press, 2007), 53–79.

Barnai, Jacob. 'Christian Messianism and the Portuguese Marranos: The Emergence of Sabbateanism in Smyrna', *Jewish History* Vol. 7, No. 2 (1993), 119–26.

_____. 'A Document from Smyrna Concerning the History of Sabbatianism' (Hebrew), *Jerusalem Studies in Jewish Thought* Vol. 2 (1982), 118–31.

_____. 'Messianism and Leadership: The Sabbatean Movement and the Leadership of the Jewish Communities in the Ottoman Empire', in Aron Rodrigue, ed., *Ottoman and Turkish Jewry: Community and Leadership* (Bloomington: Indiana University Turkish Studies, 1992), 167–82.

_____. 'Prototypes of Leadership in a Sephardic Community: Smyrna in the Seventeenth Century', in Benjamin Gampel, ed., *Crisis and Creativity in the Sephardic World 1391–1648* (New York: Columbia University Press, 1997), 146–63.

_____. 'The Sabbatean Movement in Smyrna: The Social Background', in Menachem Mor, ed., *Jewish Sects, Religious Movements, and Political Parties: Proceedings of the Third Annual Symposium of the Philip M. and Ethel Klutznick Chair in Jewish Civilization held on Sunday–Monday, October 14–15, 1990* (Omaha: Creighton University Press, 1992), 113–22.

_____. 'Some Social Aspects of the Polemics between Sabbatians and their Opponents', in Matt Goldish and Richard Popkin, eds, *Millenarianism and Messianism in Early Modern European Culture I* (Dordrecht: Kluwer Academic Publishers, 2001), 77–90.

Baron, Sabrina. 'The Guises of Dissemination in Early Seventeenth-Century England', in Brendan Dooley and Sabrina Baron, eds, *The Politics of Information in Early Modern Europe* (London: Routledge, 2001), 41–56.

Batten, J. Minton. *John Dury: Advocate of Christian Reunion* (Chicago: The University of Chicago Press, 1944).

Bearman, P., T. Bianquis, C.E. Bosworth, E. van Donzel and W.P. Heinrichs, eds. *Encyclopaedia of Islam I–XI* (Leiden: Koninklijke Brill, 2006).

Beckingham, C.F. 'The Achievements of Prester John', in Charles Beckingham and Bernard Hamilton, eds, *Prester John, the Mongols and the Ten Lost Tribes* (Aldershot: Variorum, 1996), 1–24.

Beinart, Haim and Yaacov Green, eds. *The Expulsion of the Jews from Spain* (Oxford: The Littman Library of Jewish Civilization, 2002).

_____. 'Order of the Expulsion from Spain: Antecedents, Causes, and Textual Analysis', in Benjamin Gampel, ed., *Crisis and Creativity in the Sephardic World 1391–1648* (New York: Columbia University Press, 1997), 79–94.

Ben Sasson, Hayim Hillel. 'Exile and Salvation in the Eyes of the Generation of the Spanish Diaspora' (Hebrew), in S.W. Baron, B. Dinur, S. Ettinger and I. Halpern, eds, *Yitzhak F. Baer Jubilee Volume* (Jerusalem: Merkaz Zalman Shazar, 1960), 216–27.

_____. *A History of the Jewish People* (Cambridge, Mass.: Harvard University Press, 1976).

Benite, Zvi Ben-Dor. *The Ten Lost Tribes: A World History* (Oxford: Oxford University Press, 2009).

Benton, Lauren. 'The British Atlantic in Global Context', in David Armitage and Michael Braddick, eds, *The British Atlantic World, 1500–1800* (New York: Palgrave Macmillan, 2009), 271–89.

Biale, David. 'Gershom Scholem on Jewish Messianism', in Marc Saperstein, ed., *Essential Papers on Messianic Movements and Personalities in Jewish History* (New York and London: New York University Press, 1992), 521–50.

Bittle, William. *James Nayler 1618–1660: The Quaker Indicted by Parliament* (York: William Session Ltd., 1986).

Bodian, Miriam. *Hebrews of the Portuguese Nation: Conversos and Community in Early Modern Amsterdam* (Bloomington and Indianapolis: Indiana University Press, 1997).

————. '"Men of the Nation": The Shaping of Converso Identity in Early Modern Europe', *Past and Present* No. 143 (1994), 48–76.

Brailsford, Mabel Richmond. *A Quaker from Cromwell's Army: James Nayler* (London: The Swathmore Press Ltd., 1927).

Braithwaite, William. *The Beginnings of Quakerism* (London: Macmillan and Co., 1912).

Brasz, Chaya and Yosef Kaplan, eds. *Dutch Jews as Perceived by Themselves and by Others: Proceedings of the Eighth International Symposium on the History of the Jews in the Netherlands* (Leiden: Brill, 1998).

Brown, Louise. *The Political Activities of the Baptists and Fifth Monarchy Men in England during the Interregnum* (New York: Burt Franklin, 1911).

Brownlees, Nicholas. 'Narrating Contemporaneity: Text and Structure in English News', in Brendan Dooley, ed., *The Dissemination of News and the Emergence of Contemporaneity in Early Modern Europe* (Farnham: Ashgate, 2010), 225–50.

Bruggeman, Jeroen. *Social Networks: An Introduction* (London and New York: Routledge, 2008).

Cantor, Geoffrey. *Quakers, Jews, and Science: Religious Responses to Modernity and the Sciences in Britain, 1650–1900* (Oxford: Oxford University Press, 2005).

Capp, Bernard. '*A Door of Hope* Re-opened: The Fifth Monarchy, King Charles and King Jesus', *Journal of Religious History* Vol. 32, No. 1 (2008), 16–31.

————. *The Fifth Monarchy Men: A Study in Seventeenth-Century English Millenarianism* (London: Faber and Faber, 1972).

Carlebach, Elisheva. *The Pursuit of Heresy: Rabbi Moses Hagiz and the Sabbatian Controversies* (New York: Columbia University Press, 1990).

————. 'Sabbatianism and the Jewish–Christian Polemic', *Proceedings of the Tenth World Congress of Jewish Studies, Division C, Volume II: Jewish Thought and Literature* (Jerusalem: World Union of Jewish Studies, 1990), 1–7.

Carrichio, Mario. 'News from the New Jerusalem', in Ariel Hessayon and David Finnegan, eds, *Varieties of Seventeenth and Early Eighteenth Century English Radicalism in Context* (Aldershot: Ashgate, 2009), 69–86.

Carrichio, Mario. *Religione, Politica e Commercio di Libri nella Rivolzione Inglese. Gli Autori di Giles Calvert, 1645–1653* (Genoa, 2003).

Carroll, Kenneth. 'Martha Simmonds, A Quaker Enigma', *Journal of the Friends Historical Society* Vol. 53 (1972), 31–52.

Carroll, Peter. *Puritanism and the Wilderness: The Intellectual Significance of the New England Frontier 1629–1700* (New York: Columbia University Press, 1969).

Cesarani, David, ed. *Port Jews: Jewish Communities in Cosmopolitan Maritime Trading Centres, 1550–1950* (London: Frank Cass, 2002).

Christianson, Paul. *Reformers and Babylon: English Apocalyptic Visions from the Reformation to the Eve of the Civil War* (Toronto: University of Toronto Press, 1978).

Coclanis, Peter, ed. *The Atlantic Economy during the Seventeenth and Eighteenth Centuries: Organization, Operation, Practice, and Personnel* (Columbia: University of South Carolina Press, 2005).

Cogley, Richard. 'The Ancestry of the American Indians: Thomas Thorowgood's *Iewes in America* (1650) and *Jews in America* (1660)', *English Literary Renaissance* Vol. 35, No. 2 (2005), 304–30.

_____. 'The Fall of the Ottoman Empire and the Restoration of Israel in the "Judeo-Centric" Strand of Puritan Millenarianism', *Church History* Vol. 72, No. 2 (2003), 304–32.

_____. 'Idealism vs. Materialism in the Study of Puritan Missions to the Indians', *Method and Theory in the Study of Religion* Vol. 3, No. 2 (1991), 165–82.

_____. 'John Eliot and the Millennium', *Religion and American Culture* Vol. 1, No. 2 (1991), 227–50.

_____. *John Eliot's Mission to the Indians before King Philip's War* (Cambridge, Mass.: Harvard University Press, 1999).

_____. 'John Eliot and the Origins of the American Indians', *Early American Literature* Vol. 21, No. 3 (1986/1987), 210–25.

_____. '"The Most Vile and Barbarous Nation of All the World": Giles Fletcher the Elder's "The Tarts or, Ten Tribes" (Ca. 1610)', *Renaissance Quarterly* Vol. 58, No. 3 (2005), 781–814.

_____. 'Seventeenth Century English Millenarianism', *Religion* Vol. 17 (1987), 379–96.

_____. '"Some Other Kinde of Being and Condition": The Controversy in Mid-Seventeenth-Century England over the Peopling of Ancient America', *Journal of the History of Ideas* Vol. 68, No. 1 (2007), 35–56.

Cohen, Gerson. 'Messianic Postures of Ashkenazim and Sephardim', in Marc Saperstein, ed., *Essential Papers on Messianic Movements and Personalities in Jewish History* (New York and London: New York University Press).

Constable, Olivia Remie. *Housing the Stranger in the Mediterranean World: Lodging, Trade, and Travel in Late Antiquity and the Middle Ages* (Cambridge: Cambridge University Press, 2003).

Cook, David. *Studies in Muslim Apocalyptic* (Princeton: The Darwin Press Inc., 2002).

Cooper, Frederick. 'Networks, Moral Discourse and History', in Thomas Callaghy, Ronald Kassimir and Robert Latham, eds, *Intervention and Transnationalism in Africa* (Cambridge: Cambridge University Press, 2001), 23–46.

Cressy, David. *Coming Over: Migration and Communication Between England and New England in the Seventeenth Century* (Cambridge: Cambridge University Press, 1987).

Damrosch, Leo. *The Sorrows of the Quaker Jesus: James Nayler and the Puritan Crackdown on the Free Spirit* (Cambridge, Mass.: Harvard University Press, 1996).

Davies, W.D. 'From Schweitzer to Scholem: Reflections on Sabbatai Svi', in Marc Saperstein, ed., *Essential Papers on Messianic Movements and Personalities in Jewish History* (New York and London: New York University Press, 1992), 335–74.

De Jong, James. *As the Waters Cover the Sea: Millennial Expectations in the Rise of Anglo-American Missions 1640–1810* (Kampen: J.H. Kok, 1970).

De Vivo, Filippo. *Information and Communication in Venice: Rethinking Early Modern Politics* (Oxford: Oxford University Press, 2007).

De Vries, Jan and Ad van der Woude. *The First Modern Economy: Success, Failure, and Perseverance of the Dutch Economy, 1500–1815* (Cambridge: Cambridge University Press, 1997).

Donovan, Joseph. *Pelagius and the Fifth Crusade* (Philadelphia: University of Pennsylvania Press, 1950).

Dooley, Brendan, ed. 'Introduction', *The Dissemination of News and the Emergence of Contemporaneity in Early Modern Europe* (Farnham: Ashgate, 2010), 1–19.

Dunkelgrun, Theodor. 'Neerlands Israel': Political Theology, Christian Hebraism, Biblical Antiquarianism, and Historical Myth* (Leiden: Koninklijke Brill, 2009).

Dursteler, Eric. *Venetians in Constantinople: Nation, Identity, and Coexistence in the Early Modern Mediterranean* (Baltimore: The Johns Hopkins University Press, 2006).

Eldem, Edhem. 'Foreigners at the Threshold of Felicity: The Reception of Foreigners in Ottoman Istanbul', in Donatella Calabi and Stephen Christensen, eds, *Cultural Exchange in Early Modern Europe II* (Cambridge: Cambridge University Press, 2007), 114–31.

Elior, Rachel, ed. *The Sabbatian Movement and its Aftermath: Messianism, Sabbatianism and Frankism* (Hebrew) (Jerusalem: Institute of Jewish Studies, 2001).

Emmerson, Richard Kenneth. *Antichrist in the Middle Ages: A Study of Medieval Apocalypticism, Art, and Literature* (Manchester: Manchester University Press, 1981).

Endelman, Todd. *The Jews of Britain, 1656 to 2000* (Berkeley: University of California Press, 2002).

Faroqhi, Suraiya. *The Ottoman Empire and the World Around It* (London and New York: I.B. Tauris, 2004).

————. *Pilgrims and Sultans: The Hajj under the Ottomans 1517–1683* (London: I.B. Tauris, 1996).

Fenton, Paul. 'Shabbetay Sebi and His Muslim Contemporary Muhammed an-Niyazi', in David Blumenthal, ed., *Approaches to Judaism in Medieval Times III* (Atlanta: Scholars Press, 1988), 81–8.

Filiu, Jean-Pierre. *Apocalypse in Islam*. Translated by M.B. DeBevoise (Berkeley: University of California Press, 2011).

Fogelklou, Emilia. *James Nayler: The Rebel Saint 1618–1660*. Translated by Lajla Yapp (London: Ernest Benn Ltd., 1931).

Freely, John. *The Lost Messiah: In Search of the Mystical Rabbi Sabbatai Sevi* (Woodstock: The Overlook Press, 2003).

Gampel, Benjamin, ed. *Crisis and Creativity in the Sephardic World 1391–1648* (New York: Columbia University Press, 1997).

Gelfand, Noah. 'A Transatlantic Approach to Understanding the Formation of a Jewish Community in New Netherland and New York', *New York History* Vol. 89, No. 4 (2008), 375–95.

Gibson, Kenneth. 'John Dury's Apocalyptic Thought: A Reassessment', *Journal of Ecclesiastical History* Vol. 61, No. 2 (2010), 299–313.

Glaser, Lynn. *Indians or Jews? Reprint of Manasseh ben Israel's The Hope of Israel* (Gilroy: Roy V. Boswell, 1973).

Glick, Thomas. 'On Converso and Marrano Ethnicity', Benjamin Gampel, ed., *Crisis and Creativity in the Sephardic World 1391–1648* (New York: Columbia University Press, 1997), 59–76.

Goffman, Daniel. *Izmir and the Levantine World, 1550–1650* (Seattle and London: University of Washington Press, 1990).

_____. *The Ottoman Empire and Early Modern Europe* (Cambridge: Cambridge University Press, 2002).

Goldish, Matt. *The Sabbatean Prophets* (Cambridge, Mass.: Harvard University Press, 2004).

Gow, Andrew. *The Red Jews: Antisemitism in an Apocalyptic Age 1200–1600* (Leiden: E.J. Brill, 1995).

Grafton, Anthony. *New World, Ancient Texts: The Power of Tradition and the Shock of Discovery* (Cambridge, Mass.: The Belknap Press of Harvard University Press, 1992).

Graizbord, David. *Souls in Dispute: Converso Identities in Iberia and the Jewish Diaspora, 1580–1700* (Philadelphia: University of Pennsylvania Press, 2004).

Greene, Molly. 'Beyond the Northern Invasion: The Mediterranean in the Seventeenth Century', *Past and Present* Vol. 174, No. 1 (2002), 42–71.

_____. *A Shared World: Christians and Muslims in the Early Modern Mediterranean* (Princeton: Princeton University Press, 2000).

Gross, Abraham. 'The Expulsion and the Search for the Ten Tribes', *Judaism* Vol. 41, No. 2 (1992), 130–48.

Hall, David. *The Faithful Shepherd: A History of the New England Ministry in the Seventeenth Century* (Williamson: University of North Carolina Press, 1972).

Hall, Marie Boas. *Henry Oldenburg: Shaping the Royal Society* (Oxford: Oxford University Press, 2002).

Hamilton, Bernard. 'Continental Drift: Prester John's Progress though the Indies', in Charles Beckingham and Bernard Hamilton, eds, *Prester John, the Mongols and the Ten Lost Tribes* (Aldershot: Variorum, 1996), 237–69.

————. 'The Elephant of Christ: Reynald of Chatillon', *Studies in Church History* Vol. 15 (1978), 97–104.

Hamilton, Keith and Richard Langhorne. *The Practice of Diplomacy: Its Evolution, Theory and Administration* (London and New York: Routledge, 1995).

Handover, P.M. *A History of the London Gazette 1665–1965* (London: Her Majesty's Stationery Office, 1965).

Harline, Craig. *Pamphlets, Printing, and Political Culture in the Early Dutch Republic* (Dordrecht: Martinus Nijhoff Publishers, 1987).

Hathaway, Jane. 'The Grand Vizier and the False Messiah: The Sabbatai Sevi Controversy and the Ottoman Reform in Egypt', *Journal of the American Oriental Society* Vol. 117, No. 4 (1997), 665–71.

Hedley, John, Hans Hillerbrand and Anthony Papalas, eds. *Confessionalization in Europe, 1555–1700: Essays in Honor and Memory of Bodo Nischan* (Aldershot: Ashgate, 2004).

Herzig, Tamar. 'Introduction', in Asaph Ben-Tov, Yaacov Deutsch and Tamar Herzig, eds, *Knowledge and Religion in Early Modern Europe: Studies in Honor of Michael Heyd* (Leiden: Brill, 2013), 1–12.

Heyd, Michael. 'The "Jewish Quaker": Christian Perceptions of Sabbatai Zevi as an Enthusiast', in Allison Coudert and Jeffrey Shoulson, eds, *Hebraica Veritas?: Christian Hebraists and the Study of Judaism in Early Modern Europe* (Philadelphia: University of Pennsylvania Press, 2004), 234–65.

Heywood, C.J. 'Sir Paul Rycaut, A Seventeenth-Century Observer of the Ottoman State: Notes for a Study', in Ezel Kural Shaw and C.J. Heywood, eds, *English and Continental Views of the Ottoman Empire, 1500–1800* (Los Angeles: University of California, 1972), 33–59.

Hill, Christopher. *Antichrist in Seventeenth-Century England* (London: Oxford University Press, 1971).

————. 'Till the Conversion of the Jews', in Richard Popkin, ed., *Millenarianism and Messianism in English Literature and Thought 1650–1800: Clark Library Lectures 1981–1982* (Leiden: E.J. Brill, 1988), 12–36.

Holstun, James. *A Rational Millennium: Puritan Utopias of Seventeenth-Century England and America* (New York and Oxford: Oxford University Press, 1987).

Hotson, Howard. 'Anti-Semitism, Philo-Semitism, Apocalypticism and Millenarianism in Early Modern Europe: A Case Study and Some Methodological Reflections', in Alister Chapman, John Coffey and Brad Gregory, eds, *Seeing Things Their Way: Intellectual History and the Return of Religion* (Notre Dame: University of Notre Dame Press, 2009), 91–134.

Hsia, Ronnie. *Social Discipline in the Reformation: Central Europe 1550–1750* (London: Routledge, 1991).

Huddleston, Lee Eldridge. *Origins of the American Indians: European Concepts, 1492–1729* (Austin and London: University of Texas Press, 1967).

Hull, William. *Benjamin Furly and Quakerism in Rotterdam* (Lancaster: Lancaster Press, 1941).

_____. *The Rise of Quakerism in Amsterdam, 1655–65* (Philadelphia: Patterson and White Company, 1938).

Hunter, Michael. *Boyle Between God and Science* (New Haven: Yale University Press, 2009).

_____. *Establishing the New Science: The Experience of the Early Royal Society* (Woodbridge: The Boydell Press, 1989).

_____. *The Royal Society and its Fellows 1660–1700: The Morphology of an Early Scientific Institution* (Preston: Alphaprint, 1982).

Idel, Moshe. 'Jewish Mysticism Among the Jews of Arab/Moslem Lands', *The Journal for the Study of Sephardic and Mizrahi Jewry* Vol. 1, No. 1 (2007), 14–39.

_____. *Messianic Mystics* (New Haven: Yale University Press, 1998).

_____. 'On Prophecy and Magic in Sabbateanism', *Kabbalah: Journal for the Study of Jewish Mystical Text* Vol. 8 (2003), 7–50.

_____. '"One from a Town, Two from a Clan": The Diffusion of Lurianic Kabbalah and Sabbateanism – a Re-examination', *Jewish History* Vol. 7, No. 2 (1993), 79–104.

_____. 'Religion, Thought, and Attitudes: The Impact of the Expulsion on the Jews', in Elie Kedourie, ed., *Spain and the Jews: The Sephardi Experience, 1492 and After* (London: Thames and Hudson, 1992), 123–39.

Infelise, Mario. 'The Circolazione dell'Informazione Commerciale', in Franco Franceschi, ed., *Commercio e Cultura Mercantile* (Treviso: Angelo Colle, 2007), 499–522.

_____. 'Copisti e Gazzettieri nella Venezia del Seicento', in S. Gasparri, G. Levi and P. Moro, eds, *Venezia: Itinerari per la Storia della Citta* (Bologna: Societa Editrice il Mulino, 1997), 193–219.

_____. 'From Merchants' Letters to Handwritten Political Avvisi: Notes on the Origins of Public Information', in Livio Antonielli, Carlo Capra and Mario Infelise, eds, *Cultural Exchange in Early Modern Europe III* (Cambridge: Cambridge University Press, 2007), 33–52.

_____. 'News Networks between Italy and Europe', in Brendan Dooley, ed., *The Dissemination of News and the Emergence of Contemporaneity in Early Modern Europe* (Farnham: Ashgate, 2010), 51–68.

_____. *Prima dei Giornali: Alle Origini della Pubblica Informazione* (Bari: Laterza, 2002).

_____. 'Sulle Prime Gazette a Stampa Veneziane', in Livio Antonielli, Carlo Capra and Mario Infelise, eds, *Per Marino Berengo: Studi degli Allievi* (Milan: FrancoAngeli, 2000), 469–89.

Israel, Jonathan. *Diasporas within a Diaspora: Jews, Crypto-Jews and the World Maritime Empires* (Leiden: E.J. Brill, 2002).

_____. *European Jewry in the Age of Mercantilism 1550–1750* (Oxford: The Littman Library of Jewish Civilization, 1998).

Jacobs, Martin. 'An Ex-Sabbatean's Remorse? Sambari's Polemics against Islam', *Jewish Quarterly Review* Vol. 97, No. 3 (2007), 347–78.

Janse, Wim and Barbara Pitkin. *The Formation of Clerical and Confessional Identities in Early Modern Europe* (Leiden: E.J. Brill, 2006).

Jennings, Francis. 'Goals and Functions of Puritan Missions to the Indians', *Ethnohistory* Vol. 18, No. 3 (1971), 197–212.

Johnston, Warren. *Revelation Restored: The Apocalypse in Later Seventeenth-Century England* (Woodbridge: The Boydell Press, 2011).

Kaplan, Benjamin. *Divided by Faith: Religious Conflict and the Practice of Toleration in Early Modern Europe* (London and Cambridge, Mass.: The Belknap Press of Harvard University Press, 2007).

Kaplan, Yosef. *An Alternative Path to Modernity: The Sephardic Diaspora in Western Europe* (Leiden: E.J. Brill, 2000).

_____. 'The Attitude of the Leadership of the Portuguese Community in Amsterdam to the Sabbatian Movement' (Hebrew), *Zion* Vol. 39 (1974), 198–216.

_____. *From Christianity to Judaism: The Story of Isaac Orobio de Castro* (Oxford: Oxford University Press, 1989).

_____. 'The Self-Definition of the Sephardic Jews of Western Europe and Their Relation to the Alien and the Stranger', in Benjamin Gampel, ed., *Crisis and Creativity in the Sephardic World 1391–1648* (New York: Columbia University Press, 1997), 121–45.

Kaplan, Yosef, Henry Mechoulan and Richard Popkin, eds. *Menasseh ben Israel and His World* (Leiden: E.J. Brill, 1989).

Katchen, Aaron. *Christian Hebraists and Dutch Rabbis: Seventeenth Century Apologetics and the Study of Maimonides' Mishneh Torah* (Cambridge, Mass.: Harvard University Press, 1984).

Katz, David. 'English Charity and Jewish Qualms: The Rescue of the Ashkenazi Community of Seventeenth-Century Jerusalem', in Ada Rapoport-Albert and Steven Zipperstein, eds, *Jewish History: Essays in Honour of Chimen Abramsky* (London: Peter Halban, 1988), 245–66.

_____. 'Henry Jessey and Conservative Millenarianism in Seventeenth-Century England and Holland', in Jozeph Michman, ed., *Dutch Jewish History: Proceedings of the Fourth Symposium on the History of the Jews in the Netherlands 7–10 December – Tel-Aviv-Jerusalem, 1986* (Jerusalem: 'Graf-Chen' Press, 1987), 75–93.

_____. 'Menasseh ben Israel's Mission to Queen Christina of Sweden, 1651–1655', *Jewish Social Studies* Vol. 45, No. 1 (1983), 57–72.

_____. 'Philo-Semitism in the Radical Tradition: Henry Jessey, Morgan Llwyd, and Jacob Boehme', in Johannes van den Berg and E.G. van der Wall, eds, *Jewish–Christian Relations in the Seventeenth Century: Studies and Documents* (Dordrecht: Kluwer Academic Publishers, 1988).

_____. *Philo-Semitism and the Readmission of the Jews to England 1603–1655* (Oxford: Clarendon Press, 1982).

Katz, Jacob. *Jews and Freemasons in Europe 1723–1939* (Cambridge, Mass.: Harvard University Press, 1970).

Kellenbenz, Hermann. *The Rise of the European Economy: An Economic History of Continental Europe from the Fifteenth to the Eighteenth Century* (New York: Holmes and Meier Publishers, 1976).

Klooster, Wim. 'Networks of Colonial Entrepreneurs: The Founders of the Jewish Settlements in Dutch America, 1650s and 1660s', in Richard Kagan and Philip Morgan, eds, *Atlantic Diasporas: Jews, Conversos, and Crypto-Jews in the Age of Mercantilism, 1500–1800* (Baltimore: The Johns Hopkins University Press, 2009), 33–49.

Kochan, Lionel. *The Making of Western Jewry, 1600–1819* (New York: Palgrave Macmillan, 2004).

Lankhorst, Otto. 'Newspapers in the Netherlands in the Seventeenth Century', in Sabrina Baron and Brendan Dooley, eds, *The Politics of Information in Early Modern Europe* (London: Routledge, 2001), 151–9.

Lawee, Eric. 'The Messianism of Isaac Abrabanel, Father of the [Jewish] Messianic Movements of the Sixteenth and Seventeenth Centuries', in Matt Goldish and Richard Popkin, eds, *Millenarianism and Messianism in Early Modern European Culture I* (Dordrecht: Kluwer Academic Publishers, 2001), 1–40.

Lazar, Moshe and Stephen Haliczer, eds. *The Jews of Spain and the Expulsion of 1492* (Lancaster: Labyrinthos, 1997).

Léchot, Pierre-Olivier. *Un Christianisme 'Sans Partialité': Irénisme et Méthode Chez John Dury (v. 1600–1680)* (Paris: Honoré Champion, 2011).

Liebes, Yehuda, ed. *G. Scholem, Researches in Sabbatianism* (Hebrew) (Tel Aviv: AmOved, 1991).

_____. *Studies in Jewish Myth and Jewish Messianism*. Translated by Batya Stein (New York: State University of New York Press, 1993).

Loewenstein, David. 'Scriptural Exegesis, Female Prophecy, and Radical Politics in Mary Cary', *Studies in English Literature, 1500–1900* Vol. 46, No. 1 (2006), 133–53.

Loker, Zvi. 'English Contemporary Opinions on the Sabbatean Movement', *Jewish Historical Studies* Vol. 29 (1982), 35–7.

Lowance, Mason. *The Language of Canaan: Metaphor and Symbol in New England from the Puritans to the Transcendentalists* (Cambridge, Mass.: Harvard University Press, 1980).

Lydon, Ghislaine. *On Trans-Saharan Trails: Islamic Law, Trade Networks, and Cross-Cultural Exchange in Nineteenth-Century Western Africa* (Cambridge: Cambridge University Press, 2009).

Maclear, James. 'New England and the Fifth Monarchy: The Quest for the Millennium in Early American Puritanism', *The William and Mary Quarterly* Vol. 32, No. 2 (1975), 223–60.

Magee, Gary and Andrew Thompson, eds. *Empires and Globalisation: Networks of People, Goods and Capital in the British World, c. 1850–1914* (Cambridge: Cambridge University Press, 2010).

Maier, Ingrid and Daniel Waugh. '"The Blowing of the Messiah's Trumpet": Reports about Sabbatai Sevi and Jewish Unrest in 1665–67', in Brendan Dooley, ed., *The Dissemination of News and the Emergence of Contemporaneity in Early Modern Europe* (Farnham: Ashgate, 2010), 137–54.

Manning, Patrick. *Navigating World History: Historians Create a Global Past* (New York: Palgrave Macmillan, 2003).

Marcus, Jacob. *The Colonial American Jew 1492–1776* (Detroit: Wayne State University Press, 1970).

Massey, Vera. *The Clouded Quaker Star James Nayler, 1618–1660* (York: Sessions Book Trust, 1999).

Matar, Nabil. 'The Idea of the Restoration of the Jews in English Protestant Thought, 1661–1701', *The Harvard Theological Review* Vol. 78, No. 1–2 (1985), 115–48.

_____. *Islam in Britain, 1558–1685* (Cambridge: Cambridge University Press, 1998).

Mattingly, Garrett. *Renaissance Diplomacy* (Baltimore: Penguin Books, 1964).

McGinn, Bernard. *Visions of the End: Apocalyptic Traditions in the Middle Ages* (New York: Columbia University Press, 1979).

McIntosh, Christopher. *The Rose Cross and the Age of Reason: Eighteenth-Century Rosicrucianism in Central Europe and its Relationship to the Enlightenment* (Leiden: E.J. Brill, 1992).

McKeon, Michael. 'Sabbatai Sevi in England', *Association of Jewish Studies Review* Vol. 2 (1977), 131–69.

Mechoulan, Henry and Gerard Nahon. *Menasseh ben Israel, The Hope of Israel: The English Translation by Moses Wall, 1652.* Translated by Richenda George (Oxford: Oxford University Press, 1987).

Michman, Jozeph, ed. *Dutch Jewish History: Proceedings of the Fifth Symposium of the History of the Jews in the Netherlands* (Jerusalem: Institute for Research on Dutch Jewry, 1991).

Miller, Perry. *Errand into the Wilderness* (Cambridge, Mass.: The Belknap Press of Harvard University Press, 1970).

_____. *The New England Mind: The Seventeenth Century* (Boston: Beacon Press, 1961).

Monter, William. 'Religion and Cultural Exchange, 1400–1700: Twenty-First-Century Implications', in Heinz Schilling, Istvan Toth and Robert Muchembled, eds, *Cultural Exchange in Early Modern Europe I* (Cambridge: Cambridge University Press, 2006), 3–24.

Moore, Rosemary. *The Light in their Consciences: Early Quakers in Britain 1646–1666* (University Park: The Pennsylvania State University Press, 2000).

Moore, Susan Hardman. 'New England's Reformation: "Wee shall be as a Citty upon a Hill, the Eies of all People are upon Us"', in Kenneth Fincham and Peter Lake, eds, *Religious Politics in Post-Reformation England: Essays in Honour of Nicholas Tyacke* (Woodbridge: The Boydell Press, 2006), 143–58.

Muddiman, J.G. *The King's Journalist 1659–1689: Studies in the Reign of Charles II* (New York: Augustus M. Kelley, 1971).

Nassi, Gad. 'Meliselda: The Sabbatean Metamorphosis of a Medieval Romance', *Los Muestros* Vol. 48 (2002), 38–41.

Neelon, David. *James Nayler: Revolutionary to Prophet* (Becket: Leading Press, 2009).

Nuttall, Geoffrey. *James Nayler: A Fresh Approach* (London: Friends' Historical Society, 1954).

O'Malley, Thomas. 'Religion and the Newspaper Press, 1660–1685: A Study of the *London Gazette*', in Michael Harris and Alan Lee, eds, *The Press in English Society from the Seventeenth to Nineteenth Centuries* (London and Toronto: Associated University Presses, 1986), 25–46.

Oberman, Heiko. '"Europa Afflicta:" The Reformation of the Refugees', *Archiv für Reformationsgeschichte* Vol. 83 (1992), 91–111.

Oegema, Gerbern. 'Thomas Coenen's "Ydele Verwachtinge Der Joden" (Amsterdam, 1669) as an Important Source for the History of Sabbatai Sevi', in Peter Schafer, Margarete Schluter and Giuseppe Veltri, eds, *Jewish Studies between the Disciplines: Papers in Honor of Peter Schafer on the Occasion of his Sixtieth Birthday* (Leiden: E.J. Brill, 2003), 331–53.

Parfitt, Tudor. *The Lost Tribes of Israel: The History of a Myth* (London: Weidenfeld and Nicolson, 2002).

Penman, Leigh. 'Sophistical Fancies and Mear Chimaeras? Traiano Boccalini's Ragguagli di Parnaso and the Rosicrucian Enigma', *Bruniana and Campanelliana: Ricerche Filosofiche e Materiali Storico-Testuali* (Pisa and Rome: Fabrizio Serre Editore, 2009), 101–20.

Perelis, Ronnie. '"These Indians Are Jews!": Lost Tribes, Crypto-Jews, and Jewish Self-Fashioning in Antonio de Montezinos's Relacion of 1644', in Richard Kagan and Philip Morgan, eds, *Atlantic Diasporas: Jews, Conversos, and Crypto-Jews in the Age of Mercantilism, 1500–1800* (Baltimore: The Johns Hopkins University Press, 2009), 195–212.

Perry, Micha. 'The Imaginary War Between Prester John and Eldad the Danite and its Real Implications', *Viator: Medieval and Renaissance Studies* Vol. 41, No. 1 (2010), 1–24.

Pfeiffer, Judith. 'Confessional Polarization in the 17th Century Ottoman Empire and Yusuf Ibn Ebi Abdu'd-Deyyan's *Kesfu'l-Esrar fi Ilzami'l-Yehud ve'l-Ahbar*', in Camilla Adang and Sbaine Schmidtke, eds, *Contacts and Controversies between Muslims, Jews, and Christians in the Ottoman Empire and Pre-Modern Iran* (Wurzburg: Ergon Verlag Wurzburg, 2010), 15–56.

Popkin, Richard. 'Christian Interest and Concerns about Sabbatai Zevi', in Matt Goldish and Richard Popkin, eds, *Millenarianism and Messianism in Early Modern European Culture I* (Dordrecht: Kluwer Academic Publishers, 2001), 91–101.

_____. 'Christian Jews and Jewish Christians in the 17th Century', in Richard Popkin and G.M. Weiner, eds, *Jewish Christians and Christian Jews: From the Renaissance to the Enlightenment* (Dordecht: Kluwer Academic Publishers, 1994), 57–71.

_____, ed. *Isaac La Peyrere (1596–1676): His Life, Work and Influence* (Leiden: Brill Academic Publishers, 1987).

_____. 'Jewish–Christian Relations in the Sixteenth and Seventeenth Centuries: The Conception of the Messiah', *Jewish History* Vol. 6, Nos. 1–2 (1992), 163–77.

_____. 'Millenarianism and Nationalism – A Case Study: Isaac La Peyrere', in Matt Goldish and Richard Popkin, eds, *Millenarianism and Messianism in Early Modern European Culture IV* (Dordrecht: Kluwer Academic Publishers, 2001), 77–84.

_____. 'The Rise and Fall of the Jewish Indian Theory', *Menasseh ben Israel and His World* (Leiden: E.J. Brill, 1989), 63–83.

_____. 'Seventeenth-Century Millenarianism', in Malcolm Bull, ed., *Apocalypse Theory and the Ends of the World* (Oxford: Blackwell, 1995), 112–34.

_____. 'Some Aspects of Jewish–Christian Theological Interchanges in Holland and England 1640–1700', in Johannes van den Berg and E.G. van der Wall, eds, *Jewish–Christian Relations in the Seventeenth Century: Studies and Documents* (Dordrecht: Kluwer Academic Publishers, 1988), 3–32.

_____. 'Spinoza, the Quakers and the Millenarians, 1656–1658', *Separata de Manuscrito* Vol. 6, No. 1 (1982), 113–33.

_____. *The Third Force in Seventeenth Century Thought* (Leiden: E.J. Brill, 1992).

_____. 'Three English Tellings of the Sabbatai Zevi Story', *Jewish History* Vol. 8, No. 1–2 (1994), 43–54.

Prak, Maarten. *The Dutch Republic in the Seventeenth Century: The Golden Age* (Cambridge: Cambridge University Press, 2005).

Price, J.L. *Dutch Society 1588–1713* (Harlow: Pearson Education, 2000).

Pullan, Brian. *The Jews of Europe and the Inquisition of Venice, 1550–1670* (London: I.B. Tauris, 1997).

Pullan, Brian and David Chambers, eds. *Venice: A Documentary History, 1450–1630* (Oxford: Blackwell, 1993).

Queller, Donald. *The Office of Ambassador in the Middle Ages* (Princeton: Princeton University Press, 1967).

Rapoport-Albert, Ada. *Women and the Messianic Heresy of Sabbatai Zevi 1666–1816* (Oxford: The Littman Library of Jewish Civilization, 2011).

Raymond, Joad. *The Invention of Newspaper: English Newsbooks 1641–1649* (Oxford: Clarendon Press, 2005).

_____. *Pamphlets and Pamphleteering in Early Modern Britain* (Cambridge: Cambridge University Press, 2004).

Reay, Barry. 'Popular Hostility Towards Quakers in Mid-Seventeenth-Century England', *Social History* Vol. 5, No. 3 (1980), 387–407.

_____. *The Quakers and the English Revolution* (London: Temple Smith, 1985).

Roots, Ivan. *The Great Rebellion 1642–1660* (London: B.T. Batsford Ltd., 1972).

Rosenstock, Bruce. 'Abraham Miguel Cardoso's Messianism: A Reappraisal', *Association of Jewish Studies Review* Vol. 23, No. 1 (1998), 63–104.

Rosnow, Ralph and Gary Alan Fine. *Rumour and Gossip: The Social Psychology of Hearsay* (New York: Elsevier, 1976).

Roth, Cecil. *A Life of Menasseh ben Israel: Rabbi, Printer, and Diplomat* (Philadelphia: The Jewish Publication Society of America, 1945).

———. 'The Religion of the Marranos', *The Jewish Quarterly Review* Vol. 22, No. 1 (1931), 1–33.

Rowland, Christopher. *The Open Heaven: A Study of Apocalyptic in Judaism and Early Christianity* (Eugene: Wipf and Stock, 2002).

Rozen, Minna. 'Strangers in a Strange Land: The Extraterritorial Status of Jews in Italy and the Ottoman Empire in the Sixteenth to the Eighteenth Centuries', in Aron Rodrigue, ed., *Ottoman and Turkish Jewry: Community and Leadership* (Bloomington: Indiana University Turkish Studies, 1992).

Ruderman, David. 'Hope Against Hope: Jewish and Christian Messianic Expectations in the Late Middle Ages', in Aharon Mirsky, Avrahan Grossma and Yosef Kaplan, eds, *Exile and Diaspora: Studies in the History of the Jewish People* (Jerusalem: Ben-Zvi Institute, 1991), 185–202.

Rupp, Gordon. 'Luther Against "The Turk, the Pope, and The Devil"', in Peter Brooks, ed., *Seven-Headed Luther: Essays in Commemoration of a Quincentenary 1483–1983* (Oxford: Clarendon Press, 1983), 255–73.

Saperstein, Marc, ed. *Essential Papers on Messianic Movements and Personalities in Jewish History* (New York and London: New York University Press, 1992).

Sarna, Jonathan. 'The Jews in British America', in Paolo Bernardini and Norman Fiering, eds, *The Jews and the Expansion of Europe to the West, 1450 to 1800* (New York and Oxford: Berghahn Books, 2001), 519–31.

Schacter, Jacob. 'Motivations for Radical Anti-Sabbatianism: The Case of Hakham Zevi Ashkenazi', *Sabbatianism and Its Aftermath: Messianism, Sabbatianism, and Frankism, Vol. II* (Jerusalem, 2001), 31–50.

Scholem, Gershom. 'The Crypto-Jewish Sect of the Dönmeh (Sabbatians) in Turkey', in *The Messianic Idea in Judaism and Other Essays on Jewish Spirituality* (New York: Schocken Books, 1971), 142–66.

———. 'Issac Luria: A Central Figure in Jewish Mysticism', *Bulletin of the American Academy of Arts and Sciences* Vol. 29, No. 8 (1976), 8–13.

———. *Major Trends in Jewish Mysticism* (London: Thames and Hudson, 1955).

———. *The Messianic Idea in Judaism: And Other Essays on Jewish Spirituality* (London: Schocken Books, 1971).

———. *Sabbatai Sevi: The Mystical Messiah.* Translated by R.J. Zwi Werblowsky (Princeton: Princeton University Press, 1976).

———, ed. *Studies and Texts Concerning the History of Sabbatianism and its Metamorpheses* (Hebrew) (Jerusalem: Mossad Bialik, 1982).

Schultheiss-Heinz, Sonja. 'Contemporaneity in 1672–1679: The Paris *Gazette*, the *London Gazette*, and the *Teutsche Kriegs-Kurier* (1672–1679)', in Brendan Dooley, ed., *The Dissemination of News and the Emergence of Contemporaneity in Early Modern Europe* (Farnham: Ashgate, 2010), 115–35.

Segre, Renata. 'Sephardic Refugees in Ferrara: Two Notable Families', in Benjamin Gampel, ed., *Crisis and Creativity in the Sephardic World 1391–1648* (New York: Columbia University Press, 1997), 164–85.

Setton, Kenneth. *Venice, Austria, and the Turks in the Seventeenth Century* (Philadelphia: American Philosophical Society, 1991).

Shalev, Zur. 'Islam, Eastern Christianity, and Superstition according to Some Early Modern English Observers', in Asaph Ben-Tov, Yaacov Deutsch and Tamar Herzig, eds, *Knowledge and Religion in Early Modern Europe* (Leiden: Brill, 2013), 153–80.

Sharot, Stephen. 'Jewish Millenarianism: A Comparison of Medieval Communities', *Comparative Studies in Society and History* Vol. 22, No. 3 (1980), 394–415.

_____. *Messianism, Mysticism, and Magic: A Sociological Analysis of Jewish Religious Movements* (Chapel Hill: The University of North Carolina Press, 1982).

Shmidt, Benjamin. 'The Hope of the Netherlands: Menasseh ben Israel and the Dutch Idea of America', in Paolo Bernardini and Norman Fiering, eds, *The Jews and the Expansion of Europe to the West, 1450 to 1800* (New York and Oxford: Berghahn Books, 2001), 86–106.

Silver, Abba. *A History of Messianic Speculation in Israel: From the First through the Seventeenth Centuries* (New York: The Macmillan Company, 1927).

Silverblatt, Irene. 'New Christians and New World Fears in Seventeenth-Century Peru', *Comparative Studies in Society and History* Vol. 42, No. 3 (2000), 524–46.

Sonne, Isaiah. 'New Material on Sabbatai Zevi from a Notebook of R. Abraham Rovigo' (Hebrew), *Sefunot* Vol. 3, No. 4 (1960), 39–70.

Sutcliffe, Adam. 'Jewish History in an Age of Atlanticism', in Richard Kagan and Philip Morgan, eds, *Atlantic Diasporas: Jews, Conversos, and Crypto-Jews in the Age of Mercantilism, 1500–1800* (Baltimore: The Johns Hopkins University Press, 2009), 18–30.

Sweet, Leonard. 'Christopher Columbus and the Millennial Vision of the New World', *The Catholic Historical Review* Vol. 7, No. 3 (1986), 369–82.

Terry, Altha. 'Giles Calvert's Publishing Career', *Journal of Friends Historical Society* Vol. 35 (1938), 45–9.

Tishby, Isaiah. 'Acute Apocalyptic Messianism', in Robert Seltzer, Frank Fitzgerald and Marc Saperstein, eds, *Essential Papers on Messianic Movements and Personalities in Jewish History* (New York and London: New York University Press, 1992), 259–86.

Toaff, Renzo. *La Nazione Ebrea a Livorno e a Pisa, 1591–1700* (Florence: Leo S. Olschki Editore, 1990).

Trivellato, Francesca. *The Familiarity of Strangers: The Sephardic Diaspora, Livorno, and Cross-Cultural Trade in the Early Modern Period* (New Haven and London: Yale University Press, 2009).

_____. 'Merchants' Letters across Geographical and Social Boundaries', in Francisco Bethencourt and Florike Egmond, eds, *Cultural Exchange in Early Modern Europe III* (Cambridge: Cambridge University Press, 2007), 80–103.

_____. 'Sephardic Merchants in the Early Modern Atlantic and Beyond: Towards a Comparative Historical Approach to Business Cooperation', in Richard L. Kagan and Philip D. Morgan, eds, *Atlantic Diasporas: Jews, Conversos, and Crypto-Jews in the Age of Mercantilism, 1500–1800* (Baltimore: The Johns Hopkins University Press, 2009), 99–120.

'A Turkish View of Quakerism, 1659', *Journal of the Friends Historical Society* Vol. 8 (1911), 25–7.

Van der Wall, Ernestine. 'The Amsterdam Millenarian Petrus Serrarius (1600–1669) and the Anglo-Dutch Circle of Philo-Judaists', in Johannes van den Berg and E.G. van der Wall, eds, *Jewish–Christian Relations in the Seventeenth Century: Studies and Documents* (Dordrecht: Kluwer Academic Publishers, 1988), 73–94.

_____. 'Mystical Millenarianism in the Early Modern Dutch Republic', in John Laursen and Richard Popkin, eds, *Millenarianism and Messianism in Early Modern European Culture IV* (Dordrecht: Kluwer Academic Publishers, 2001), 37–47.

_____. *De Mystieke Chiliast Petrus Serrarius (1600–1669) en Zijn Wereld* (Leiden: I.G.C. Printing, 1987).

_____. 'Three Letters by Menasseh ben Israel to John Durie: English Philo-Judaism and the Spes Israelis', *Nederlands Archief voor Kerkgeschiedenis* Vol. 65 (1985), 46–63.

Van Koningsveld, P.S., J. Sadan and Q. Al-Samarrai. *Yemenite Authorities and Jewish Messianism: Ahmad ibn Nasir al-Zaydi's Account of the Sabbathian Movement in Seventeenth Century Yemen and its Aftermath* (Leiden: Documentatiebureau Islam-Christendom, 1990).

Van Wijk, Jetteke. 'The Rise and Fall of Shabbatai Zevi as Reflected in Contemporary Press Reports', *Studia Rosenthaliana* Vol. 33 (1999), 7–27.

Veluwekamp, Jan Willem. 'International Business Communication Patterns in the Dutch Commercial System, 1500–1800', in Hans Cools, Marika Keblusek and Badelock Noldus, eds, *Your Humble Servant: Agents in Early Modern Europe* (Hilversum: Uitgeverij Verloren, 2006).

Villani, Stefano. 'Conscience and Convention: The Young Furly and the Hat Controversy', in Sarah Hutton, ed., *Benjamin Furly 1646–1714: A Quaker Merchant and His Milieu* (Florence: Leo. S. Olschki Editore, 2007), 87–110.

_____. 'Un Masaniello Quacchero: James Nayler', *Rivista di Storia e Letteratura Religiosa* (Florence: Leo. S. Olschki, 1997), 67–91.

_____. *I Primi Quaccheri e gli Ebrei* (Rome: Edizioni di Storia e Letteratura, 1997).

_____. 'Seventeenth-Century Italy and English Radical Movements', in Ariel Hessayon and David Finnegan, eds, *Varieties of Seventeenth- and Early Eighteenth-Century English Radicalism in Context* (Farnham: Ashgate, 2011), 145–60.

Voss, Rebekka. 'Entangled Stories: The Red Jews in Premodern Yiddish and German Apocalyptic Lore', *Association for Jewish Studies Review* Vol. 36, No. 1 (2012), 1–41.

Willard, Thomas. 'The Rosicrucian Manifestos in Britain', *Bibliographical Society of America Papers* Vol. 77 (1983), 489–95.

Winship, Michael. *Making Heretics: Militant Protestantism and Free Grace in Massachusetts, 1636–1641* (Princeton and Oxford: Princeton University Press, 2002).

Yates, Frances. *The Rosicrucian Enlightenment* (London and New York: Routledge, 1972).

Yerushalmi, Yosef Hayim. 'Exile and Expulsion in Jewish History', in Benjamin Gampel, ed., *Crisis and Creativity in the Sephardic World 1391–1648* (New York: Columbia University Press, 1997), 3–22.

_____. *From Spanish Court to Italian Ghetto: Isaac Cardoso: A Study in Seventeenth-Century Marranism and Jewish Apologetics* (Seattle and London: University of Washington Press, 1981).

Yovel, Yirmiyahu. *The Other Within, The Marranos: Split Identity and Emerging Modernity* (Princeton and Oxford: Princeton University Press, 2009).

Zahedieh, Nuala. 'Making Mercantilism Work: London Merchants and Atlantic Trade in the Seventeenth Century', *Transactions of the Royal Historical Society* Vol. 9 (1999), 143–58.

Zakai, Avihu. *Exile and Kingdom: History and Apocalypse in the Puritan Migration to America* (Cambridge: Cambridge University Press, 1992).

Unpublished Theses

Higgins, Lesley Hall. 'Radical Puritans and Jews in England, 1648–1672'. Doctoral dissertation, Yale University, 1967.

Terry, Altha. 'Giles Calvert, Mid-Seventeenth Century English Bookseller and Publisher: An Account of his Publishing Career, with a Checklist of his Imprints'. Master thesis, Columbia University, 1937.

Van Wijk, N.H. 'Wachtend op. de Wolk naar Jeruzalem: De Verslaglegging Rond Shabbatai Tsvi in Nederlandse Pamfletten en Couranten'. Doctoral dissertation, University of Amsterdam, 1996.

Index